First published in 1998 by
Stanley Thornes (Publishers) Ltd
Ellenborough House
Wellington Street
CHELTENHAM GL50 1YW

99 00 01 02 / 10 9 8 7 6 5 4

A catalogue record for this book is available from the British Library.

ISBN 0 7487 2975 5

Original design concept by Studio Dorel
Cover design by John Christopher, Design Works
Artwork by Maltings Partnership and Peters and Zabransky
Cartoons by Clinton Banbury
Typeset by Tech Set Ltd
Printed and bound in Italy by G. Canale & C.S.p.A., Borgaro T.se, Turin

Acknowledgements

The publishers thank the following for permission to reproduce copyright material:
British Museum, p. 239; Martyn Chillmaid, pp. 2, 9, 34, 42, 81, 89 (book), 134, 135, 136, 137 (tetrahedron, wizard's hat, cornetto, traffic cone), 138 (tetrahedron, toblerone, oxo, brick), 141, 142, 157, 162, 164, 193, 196, 253, 303; Geoscience Features Picture Library – The 'Castle', Chichen Itza, p. 137; Images, p. 31; Mary Evans Picture Library, p. 151; Paul Smith, p. 161; Quadrant, M. Simon Matthews (railway lines), p. 89; Quadrant, Tony Hobbs – Louvre, Paris, p. 137; Science and Society Picture Library, p. 203.
All other photographs STP Archive.

The publishers have made every effort to contact copyright holders but apologise if any have been overlooked.

1 Statistics: about our school

CORE

QUESTIONS

HELP YOURSELF

Addition 1

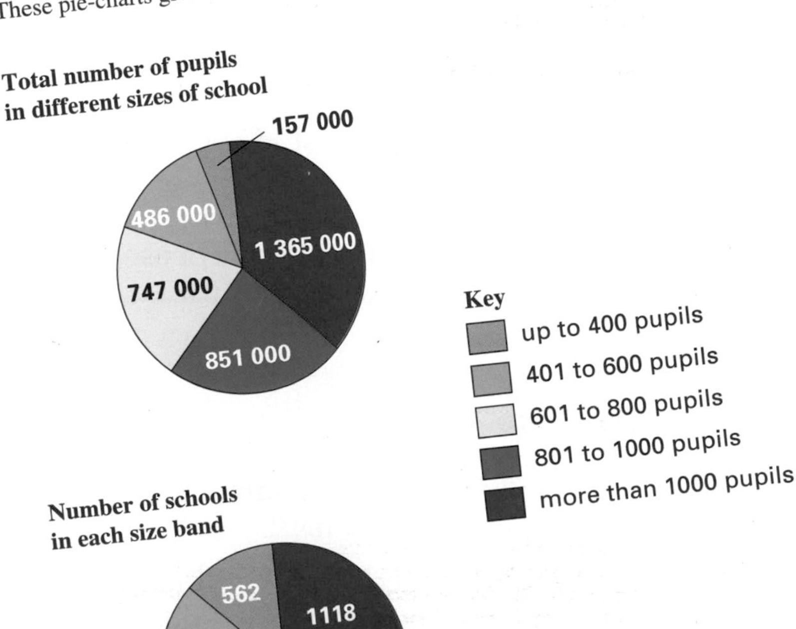

1 Surveys

In a **survey** we collect information. This information is also called **data**. The data can be sorted and shown on a diagram.

Statistics — The study of facts about numbers is called **statistics**.

Bar-chart — A **bar-chart** is a diagram made up of bars. Each bar **represents** part of the data.

Exercise 1:1

W1 **1** Sarah has done a survey of how her class travel to school. Here is her data:

walk, car, car, bus, bus, bike, walk, walk, car,
train, bus, bike, walk, walk, walk, bus, car,
train, bus, bus, walk, walk, bus, walk, bike

Sarah sorts her data:

car, car, car, car
bus, bus, bus, bus, bus, bus, bus,
train, train
bike, bike, bike
walk, walk, walk, walk, walk, walk, walk, walk, walk

Sarah counts the number of trains. There are two trains. Sarah colours in the first two squares to make the train bar.

a Complete the train bar for Sarah's class on your worksheet.

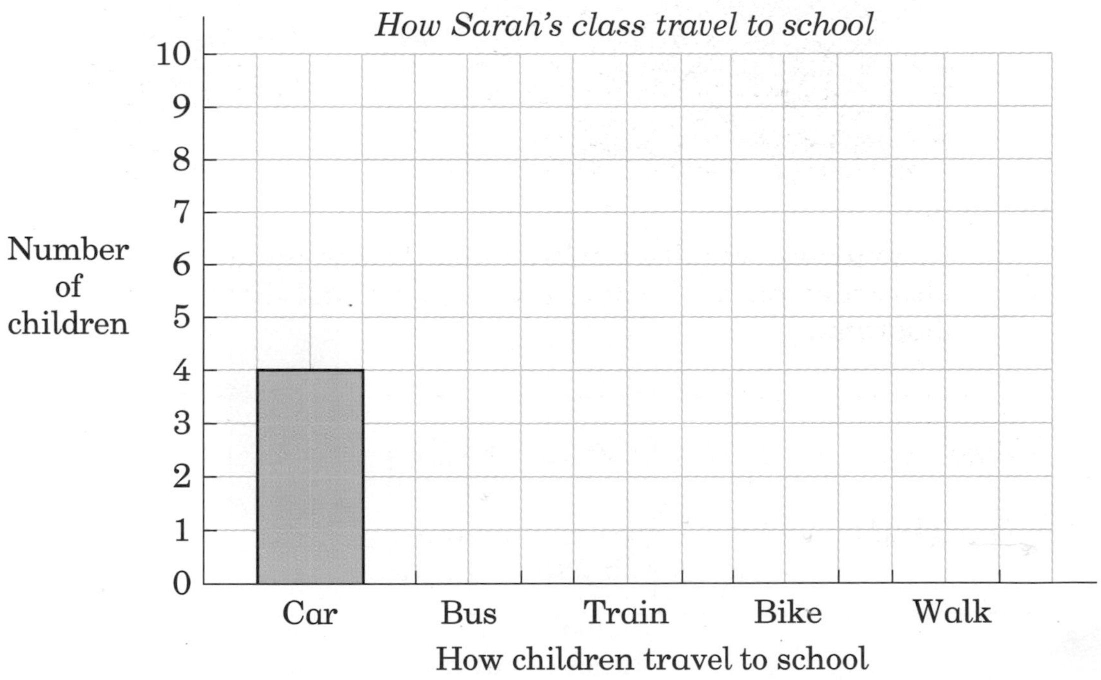

b How many children travel to school by car?

c How many children travel to school by bus?

d How many children travel to school by train?

e How do most children travel to school?

H 1 **2** **a** Do a survey like Sarah's to find out how your class travel to school.

b Make a bar-chart of your data. Remember to give your bar-chart a title.

c How many children travel to school by car?

d How many children travel to school by bus?

e How many children travel to school by bike?

f How do most children travel to your school?

3 Andrew has done a survey of how his class have their lunch. This is a bar-chart of his data:

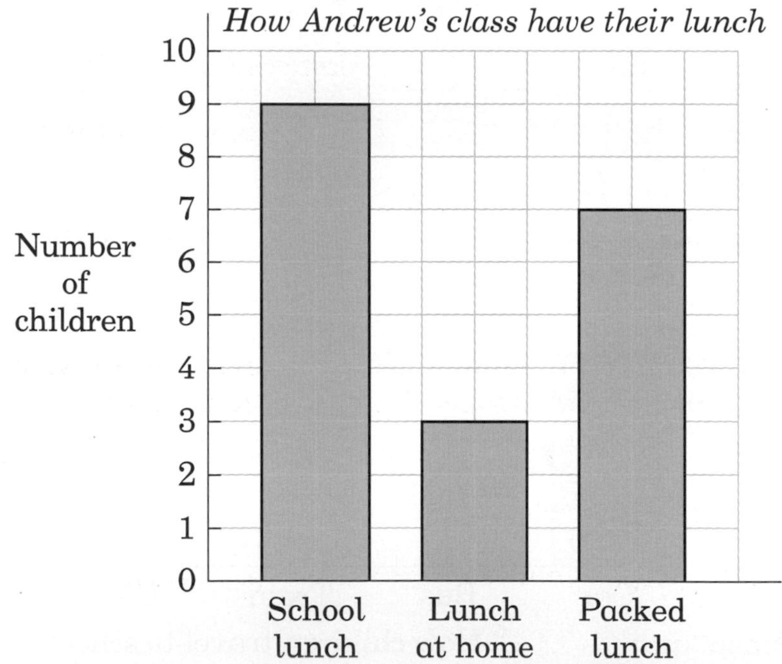

a How many children have a packed lunch?

b How many children have lunch at home?

c How many children have school lunch?

d How many children are there in Andrew's class? (Count how high each bar is.)

4 **a** Do a survey like Andrew's of how your class have their lunch.

b Make a bar-chart of your data.

c How many children in your class have a packed lunch?

d How many children in your class have lunch at home?

e How many children in your class have school lunch?

f How many children are there in your class?

Pictogram	A **pictogram** is a diagram which uses pictures instead of bars.
Key	A pictogram must always have a **key** to show what each small picture represents.
Example	Key: 웃 represents 1 child. 웃 웃 웃 웃 represents 4 children. $1 + 1 + 1 + 1 = 4$

---

5 Here is Andrew's class again. This time he has drawn a pictogram to show how his class have their lunch.

Key: Ψ represents 1 child.

How Andrew's class have their lunch

School lunch	ΨΨΨΨΨΨΨΨΨ
Lunch at home	ΨΨΨ
Packed lunch	ΨΨΨΨΨΨΨ

a How many children have a packed lunch?

b How many children have a school lunch?

c How do most children have their lunch?

6 Here is Sarah's class again. This time she has started to draw a pictogram to show how her class travel to school.

Key: 🯅 represents 1 child.

How Sarah's class travel to school

Car 🯅 🯅 🯅 🯅

Complete her pictogram in your book.

7 children travel by bus
2 children travel by train
3 children travel by bike
9 children walk to school

7 Janice did a survey on the eye colour of the children in her class. She has drawn a pictogram of her data.

Key: 👁 represents 1 child.

The eye colour of children in Janice's class

Blue	👁 👁 👁 👁 👁 👁 👁 👁 👁
Green	👁 👁 👁
Brown	👁 👁 👁 👁 👁 👁 👁 👁
Grey	👁 👁 👁

a How many children have brown eyes?

b How many children have green eyes?

c Which colour eyes do most children have?

d How many children are there in Janice's class?

W 2

8 **a** Do a survey like Janice's of the eye colour of children in your class.

b Make a pictogram of your data.

c Is your pictogram the same as Janice's? Write a sentence to say what is different.

G 1, 2, 3

2 Pie-charts

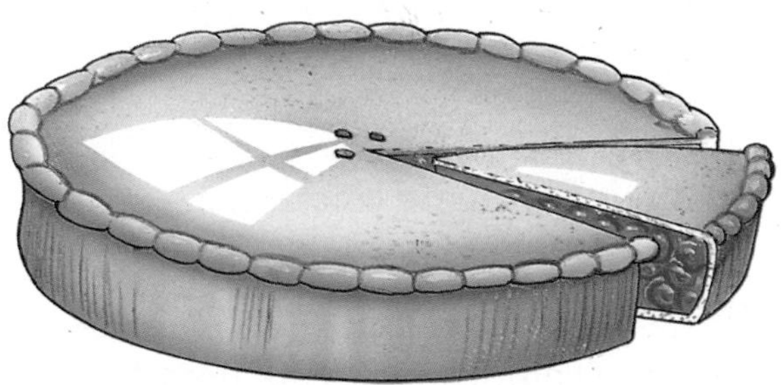

Pie-chart Another type of diagram is a **pie-chart**.

Exercise 1:2

1 You are going to make a pie-chart.

a On a sheet of paper draw a large circle.
Cut out the circle.

b (1) Fold the circle in half.

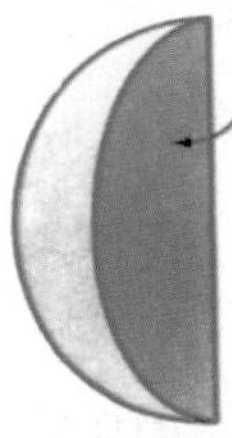

(2) Then fold it in half again.

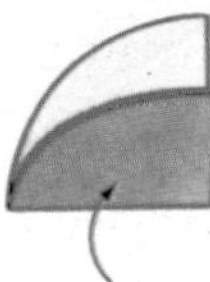

(3) Then fold it in half again.

c Unfold your circle.
It should have 8 slices.

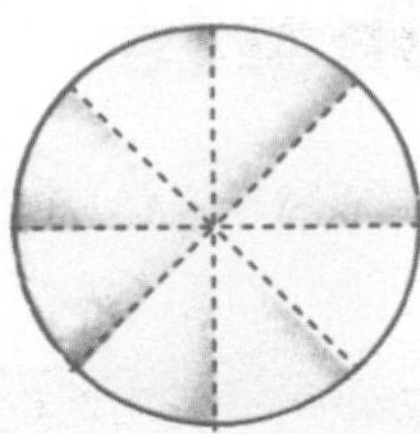

d Ask 8 children what their favourite colour is.
Write down your data.

e Colour the slices of the circle to match the favourite colours.
If two slices are the same colour, put these next to each other.

f Glue your pie-chart next to your data.
Give your pie-chart a title:
Favourite colours of 8 children in my class

2 Peter has done a survey of the eye colour of 8 children in his class.
Here is his pie-chart:

Eye colour of 8 children in Peter's class

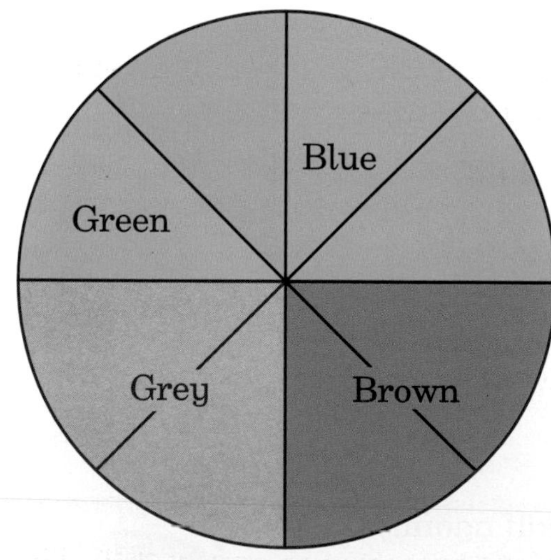

a How many children have brown eyes?

b How many children have green eyes?

c How many children have blue eyes?

d What colour eyes do most children have?

3 Tallying

These children are doing a survey of how children travel to school.
They are using a **tally-table**.

When Sarah did her survey of how her class travel to school she wrote her data like this:

walk, car, car, bus, bus, bike, walk, walk, car,
train, bus, bike, walk, walk, walk, bus, car,
train, bus, bus, walk, walk, bus, walk, bike

Martin said that a tally-table would be easier to use.

Tally marks — **Tally marks** are drawn in groups of 5. The fifth tally mark goes across the other 4.

Example

This is three |||
This is four ||||
This is five ~~||||~~

Martin put Sarah's data into a tally-table.

How Sarah's class travel to school	Tally	Total
Train	||	2
Car	||||	4
Bus	𝍸 ||	7
Bike	|||	3
Walk	𝍸 ||||	9
	Total	

Exercise 1:3

1 Here are some tally marks.
Copy them into your book.
Write down the numbers they represent.

a 𝍸 𝍸 𝍸 ||

b 𝍸 |||

c 𝍸 𝍸 ||

d 𝍸 ||

 3

2 Adam has done a survey of favourite pets.
Here is his data:

cat, dog, cat, fish, bird, dog, dog, dog,
rabbit, fish, cat, dog, cat, fish, dog, bird,
dog, dog, cat, dog, bird, rabbit, bird,
cat, bird, dog, cat

 1, 4, 5

a Put Adam's data into this tally-table:

Favourite pet	Tally	Total
Cat		
Dog		
Fish		
Bird		
Rabbit		
	Total	

b Draw a bar-chart of your data on a chart like this.

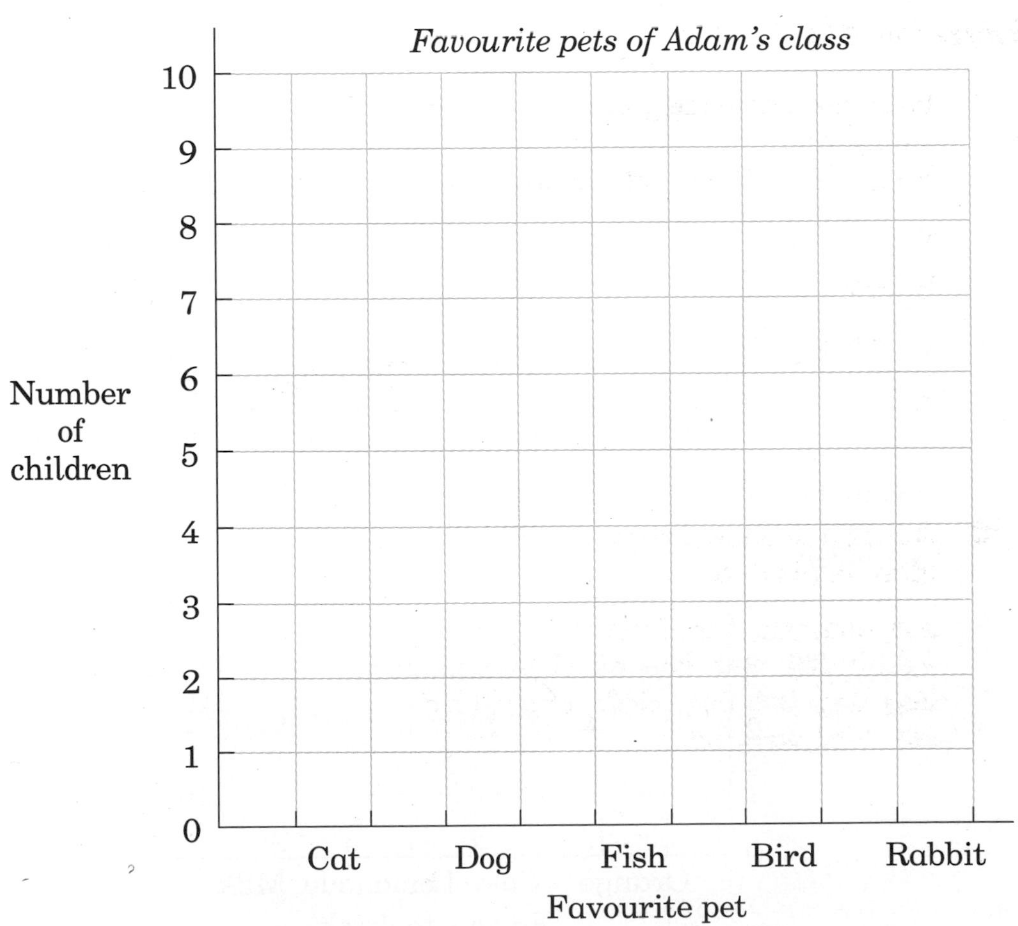

W 4 **3** Eve has done a survey of favourite drinks.
Here is her data:

orange, cola, lemonade, lemonade, lemonade, milk, cola, orange, lemonade, cola, cola, cola, milk, cola, cola, lemonade, orange, orange, orange, cola, cola

a Put Eve's data into this tally-table:

Favourite drinks	Tally	Total
Orange		
Cola		
Lemonade		
Milk		
	Total	

b Draw a bar-chart of your data on a chart like this.

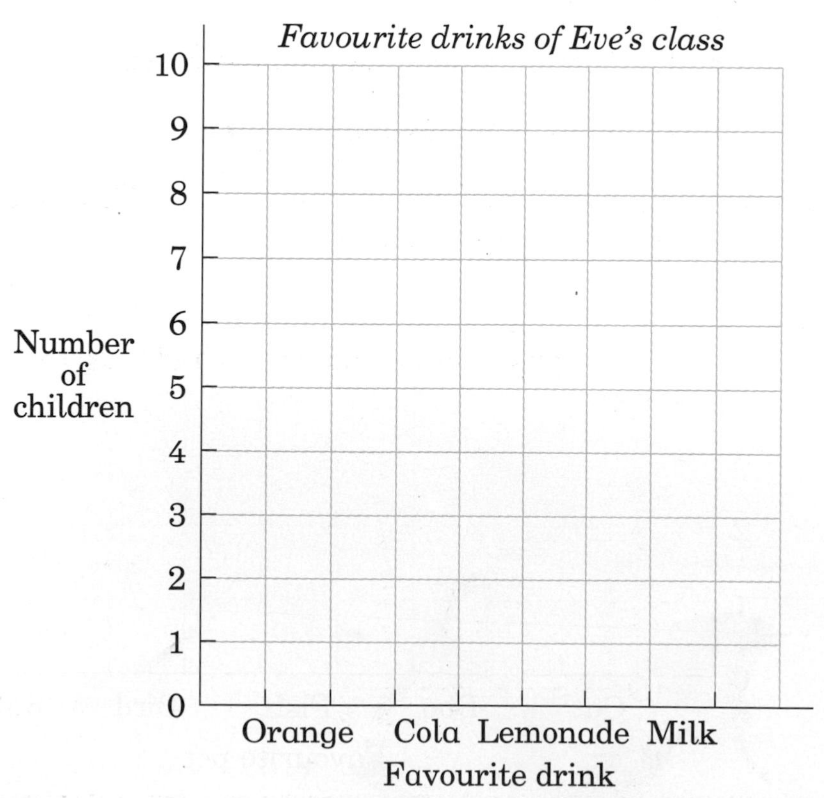

4 Location

Tariq is using co-ordinates on his map to help him find his way to the campsite.

Axes We draw a horizontal line and a vertical line known as **axes**.

***x* axis** The horizontal line is called the ***x* axis**.

***y* axis** The vertical line is called the ***y* axis**.

Co-ordinates We use two numbers to mark a point. These numbers are called **co-ordinates**.

Co-ordinates are written like this: (3, 2)
This means 3 **along** the ***x* axis** and 2 **up** the ***y* axis**.

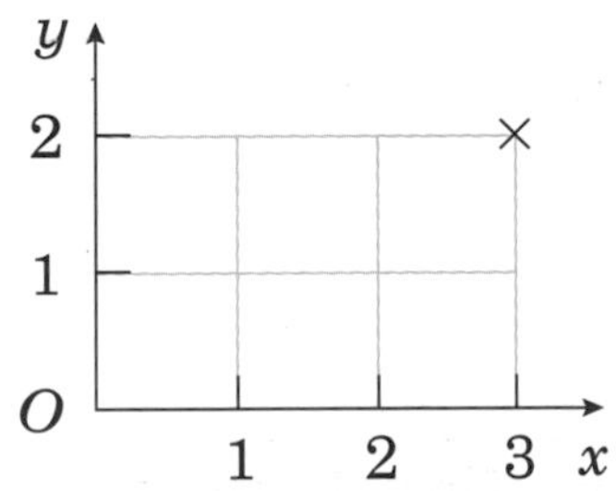

Exercise 1:4

1 This is a map of Grid Island:

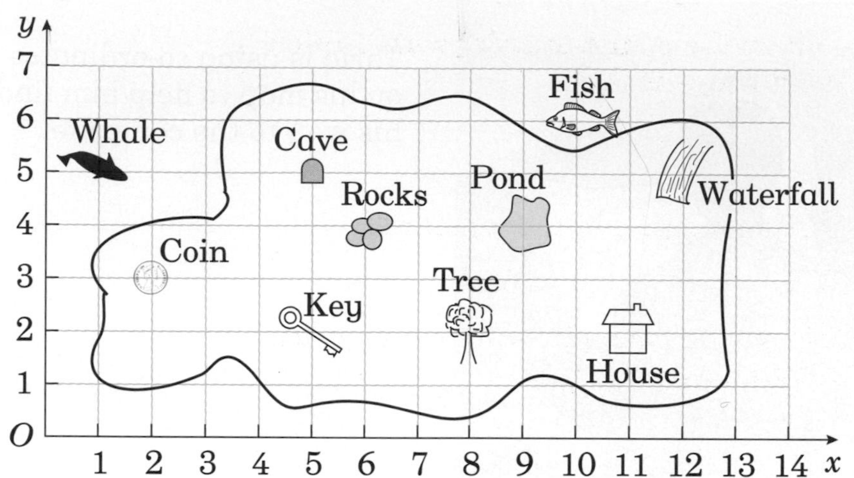

Write down the co-ordinates of each of the objects.
The first one is done for you.

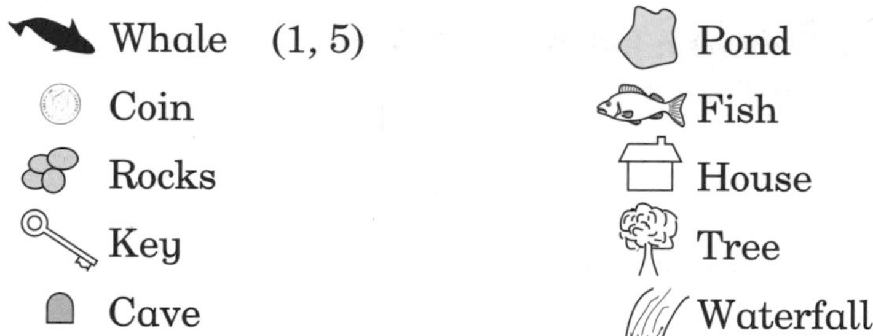

G 6, 7

W 5 You need Worksheet 5 for Questions **2** to **8**.

2 Fill in the missing numbers on each grid.

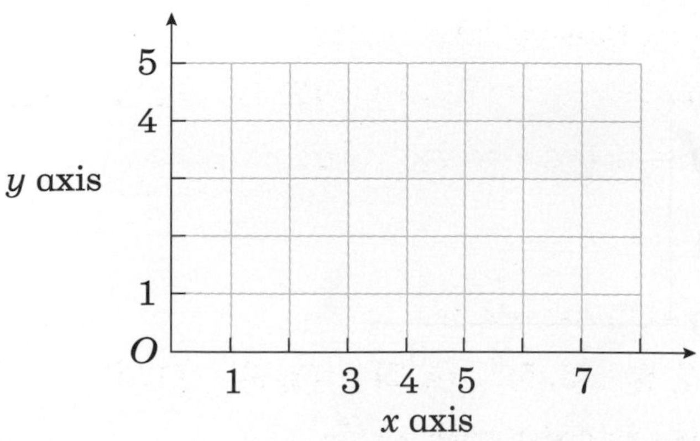

3 On your first grid plot these points.
Remember to go along the x axis and up the y axis.

(1, 3) (5, 3) (5, 6) (1, 6)

Join up the points in order with a ruler.
What shape is this?

4 On a new grid plot these points.

(3, 8) (1, 6) (3, 0) (5, 6)

Join up the points in order with a ruler.
What shape is this?

5 On a new grid plot these points.

(3, 2) (5, 8) (1, 8)

Join up the points in order with a ruler.
What shape is this?

6 On a new grid plot these points.

(1, 0) (5, 0) (5, 4) (1, 4)

Join up the points in order with a ruler.
What shape is this?

7 On a new grid plot these points.

(1, 2) (2, 3) (4, 1) (6, 4) (4, 6) (2, 4) (1, 5) (1, 2)

Join up the points in order with a ruler.

8 On a new grid plot these points.

(1, 5) (4, 0) (7, 5) (5, 5) (4, 7) (3, 5) (1, 5)

Join up the points in order with a ruler.

1 Vicky has done a survey of children's favourite animals in her class.
Here is her bar-chart.

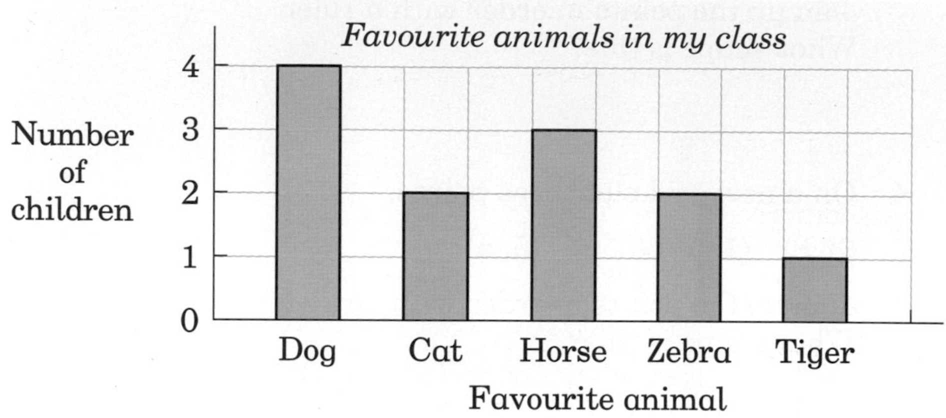

a How many children said dog?

b How many children said tiger?

c How many children said cat?

d What did most children say?

W 6 **2** David has done a survey of favourite animals in his class.
Here is his data.

Dog	3
Cat	6
Horse	4
Zebra	1
Rabbit	7
Tiger	2

Draw a bar-chart of these results.

3 Jane has done a survey of children's favourite fruits.
Here is her data:

Key: 🯅 represents 1 child.

Favourite fruits in Jane's class

Strawberry	🯅 🯅 🯅 🯅
Banana	🯅 🯅
Pear	🯅 🯅 🯅 🯅 🯅 🯅
Apple	🯅
Peach	🯅 🯅 🯅 🯅 🯅

a How many children said banana?

b How many children said peach?

c How many children said strawberry?

d What did most children say?

4 Wasim has done a survey of the favourite fruits of 8 children in his class.
Here is his data:

Favourite fruits in Wasim's class

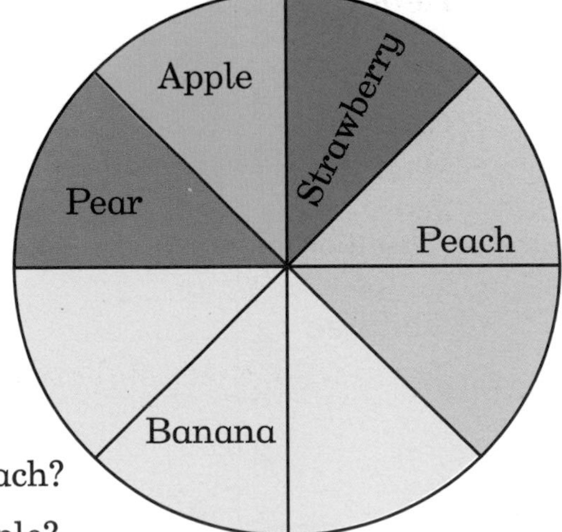

a How many children said peach?

b How many children said apple?

c How many children said pear?

d What did most children say?

W 6 **5** Janet has done a survey of favourite pets.
Here is her data:

cat, dog, fish, cat, cat, dog, dog, fish,
bird, rabbit, cat, dog, dog, dog, dog,
rabbit, hamster, hamster, rabbit, fish, fish,
bird, dog, cat, dog, hamster, rabbit

a Put Janet's data into this tally-table:

Favourite pet	Tally	Total
Cat		
Dog		
Fish		
Bird		
Rabbit		
Hamster		
	Total	

b Copy the grid below.
Draw a bar-chart of Janet's data.

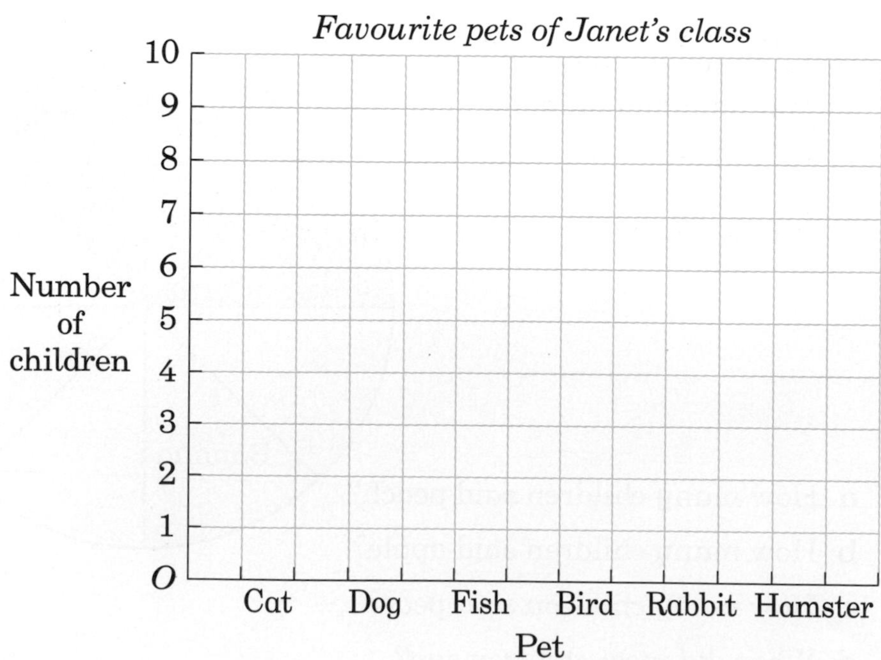

W 6 **6** You need a grid like this:

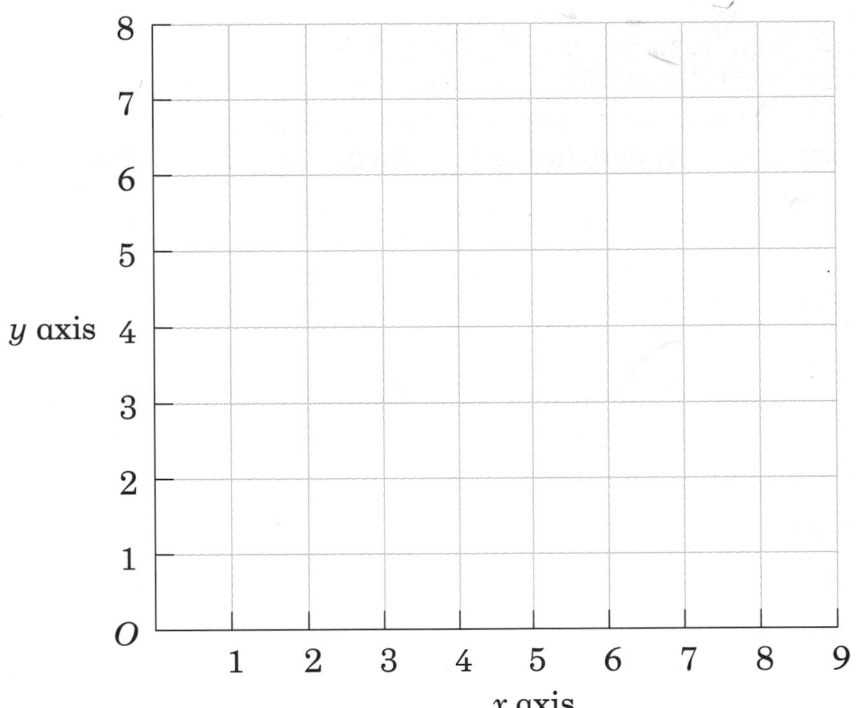

a Plot these points:

(2, 1) → (2, 6) → (1, 6)
→ (2, 7) → (6, 7) → (6, 8)
→ (7, 8) → (7, 7) → (8, 7)
→ (9, 6) → (8, 6) → (8, 1)

b Join up the points in order with a ruler.

Addition 1

Copy these into your book.
Work out the answers.

Exercise 1

1 $6 + 4$

2 $7 + 5$

3 $8 + 7$

4 $9 + 5$

5 $6 + 5$

6 $8 + 4$

7 $9 + 7$

8 $6 + 8$

Exercise 2

1 $13 + 3$

2 $17 + 2$

3 $11 + 8$

4 $17 + 2$

5 $12 + 5$

6 $18 + 2$

7 $12 + 4$

8 $13 + 4$

Exercise 3

1 $12 + 6$

2 $14 + 4$

3 $12 + 6$

4 $14 + 6$

5 $13 + 5$

6 $15 + 2$

7 $14 + 3$

8 $16 + 2$

Exercise 4

1 $13 + 7$

2 $15 + 4$

3 $12 + 7$

4 $15 + 3$

5 $16 + 3$

6 $12 + 8$

7 $13 + 6$

8 $17 + 3$

2 Symmetry

CORE

1 **Lines of symmetry**

2 **Symmetry in real life**

3 **Turnings**

QUESTIONS

HELP YOURSELF

Subtraction 1

1 Lines of symmetry

This picture of a house has a line of symmetry.
It can be split down the middle so that one half is a reflection of the other.

Line of symmetry

A **line of symmetry** divides a shape into two equal parts.
Each part is a reflection of the other.
If you fold the shape along the line of symmetry each part fits exactly on top of the other.

You can test for symmetry using a mirror.
Place a mirror on the dotted line:

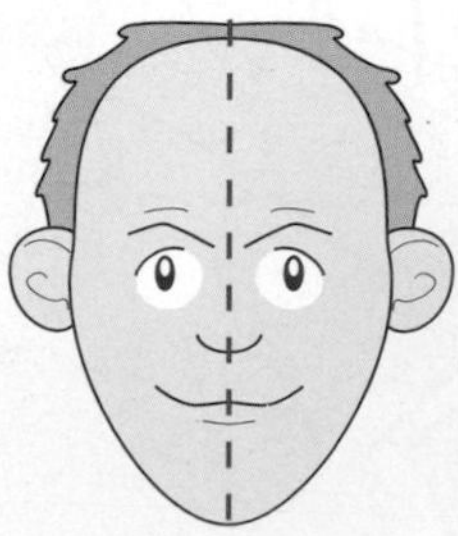

You can see that the shape in the mirror is the same as the shape in the book.
The dotted line is a line of symmetry.

Exercise 2:1

1 Which of these shapes have line symmetry?
Use your mirror to help you.

a

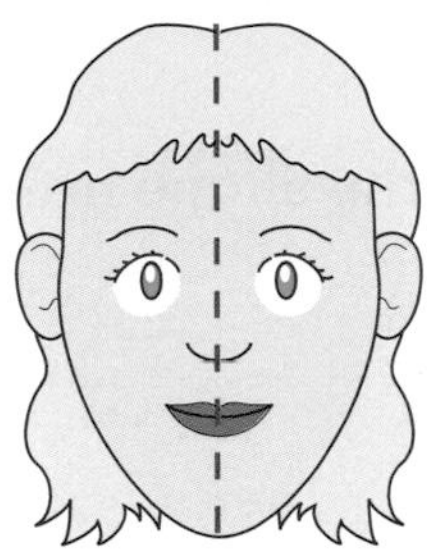

e

b

f

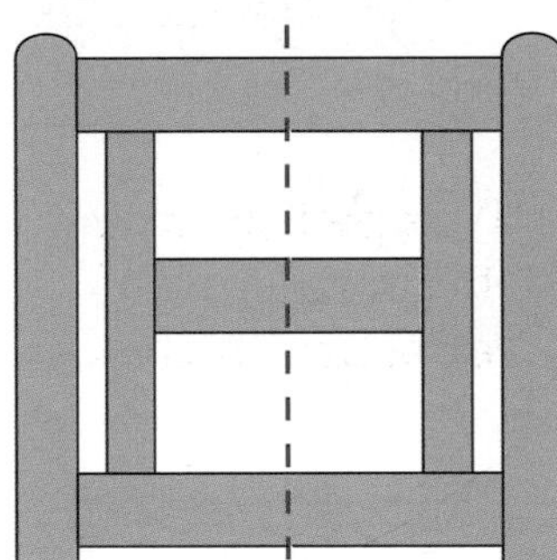

c

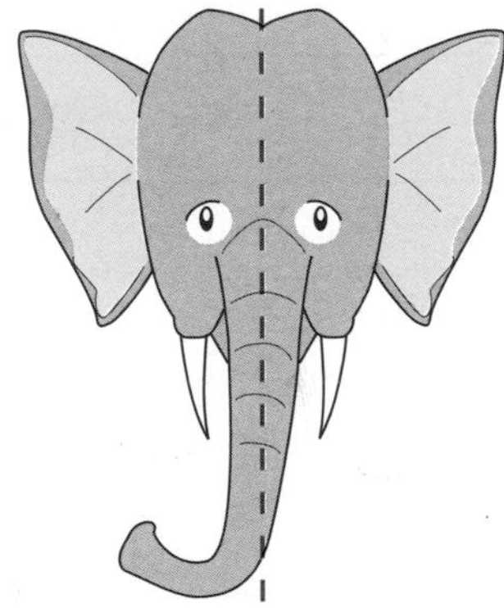

g

d

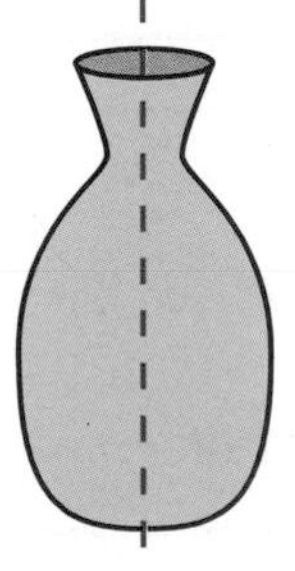

h

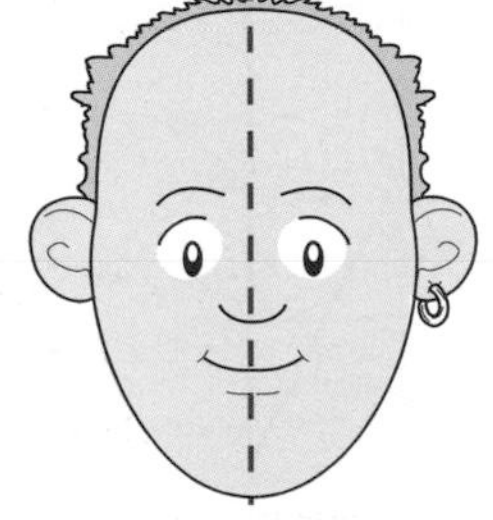

Exercise 2:2

W 1 **1** You will need copies of the shapes below.

a Cut out the shapes.

b Fold them along the dotted line.

c If the two halves match exactly then the shape has a line of symmetry.

d Stick the shapes into your book.

(1)

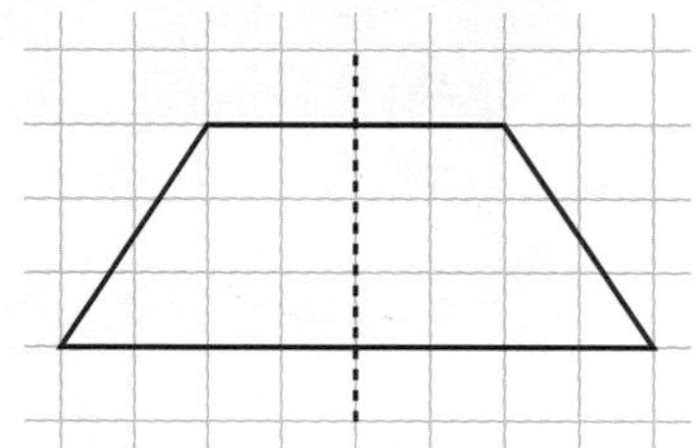

(4)

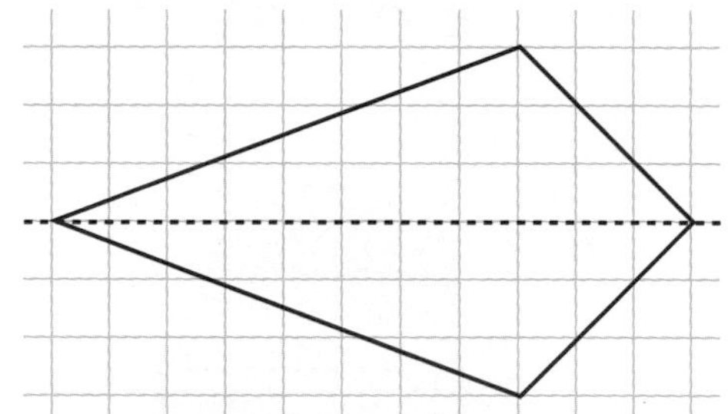

(2)

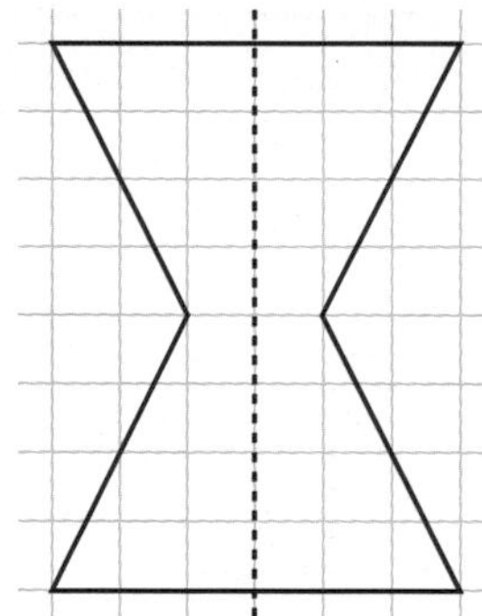

(5)

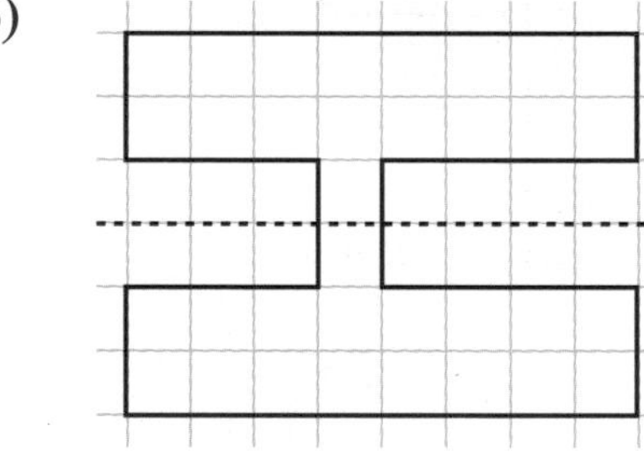

(3)

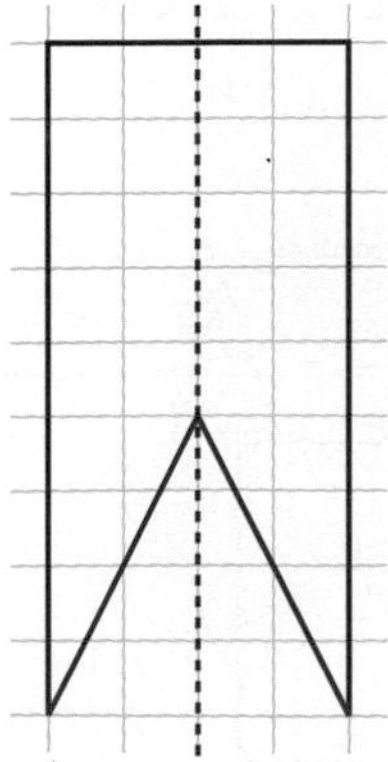

(6)

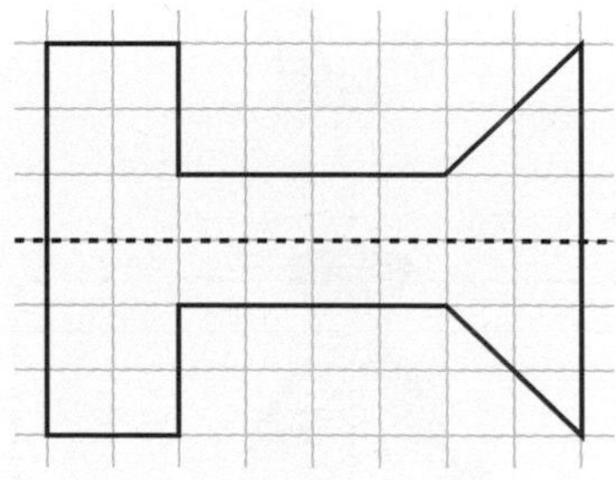

Exercise 2:3

1 You can test to see if a shape has a line of symmetry using tracing paper.
Trace each shape.
Now turn your tracing paper over.
If the picture is the same, then it has a line of symmetry.

a

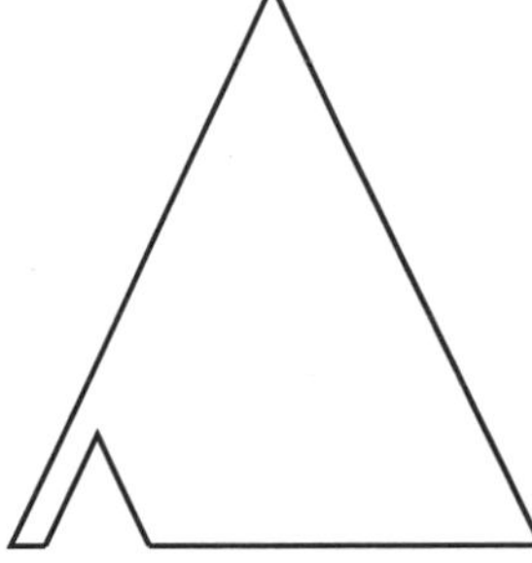

e

b

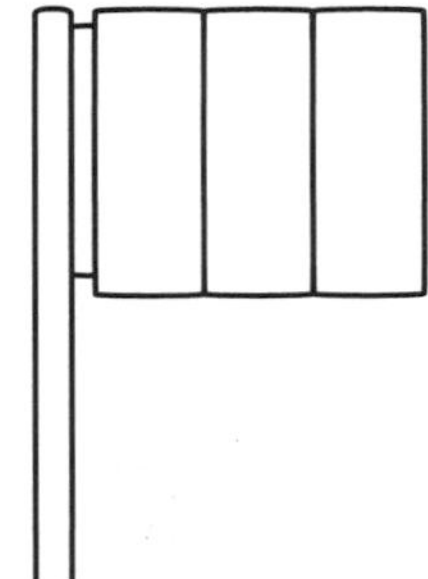

f

c

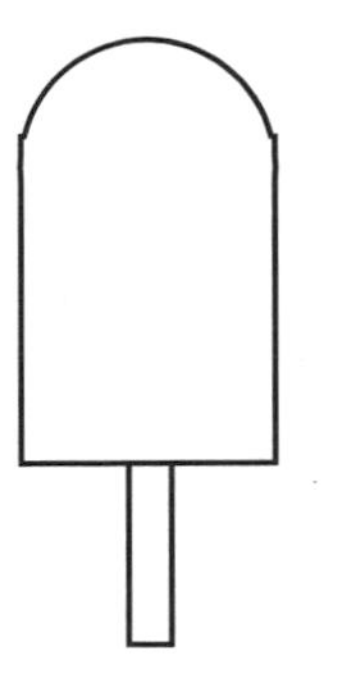

g

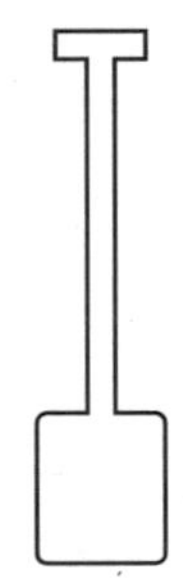

d

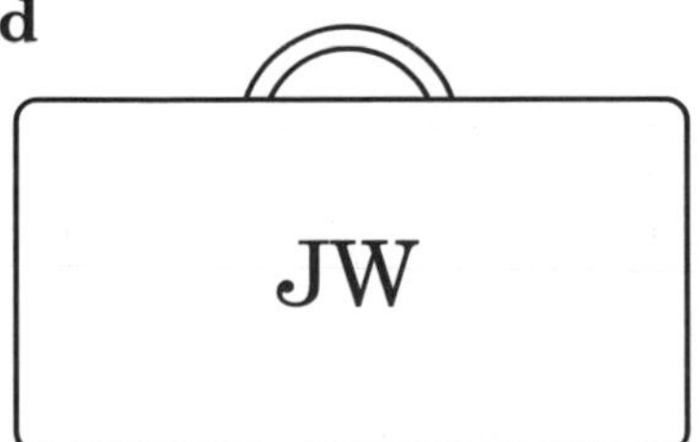

h

Exercise 2:4

W 2 **1** You can only see half of this shape:

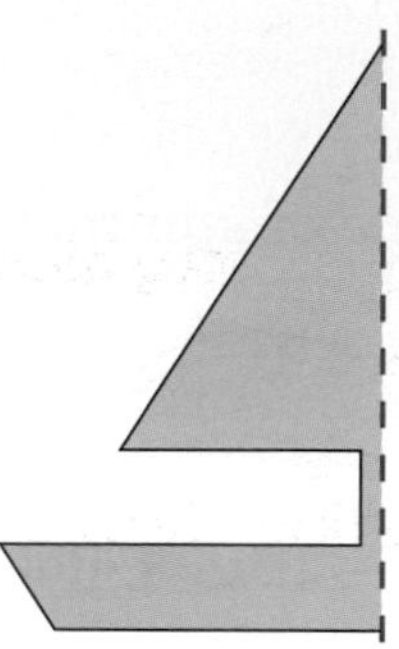

a Put a mirror on the dotted line.
You can now see all of the shape.

b On your worksheet, fold along the line of symmetry (the dotted line).

c Cut out the shape.

It now looks like this:

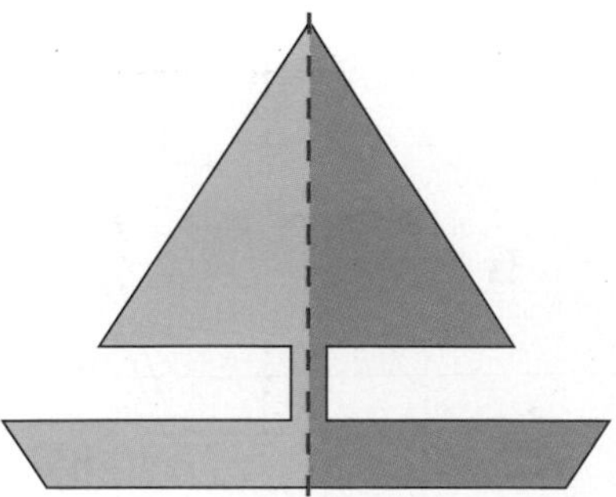

H 1 **d** Complete the other shapes on your worksheet and stick them into your book.

Exercise 2:5

1 a On squared paper draw a rectangle 8 cm by 6 cm.
Cut it out.

b Fold the rectangle in half.
Unfold your shape.

c Using a dotted line, draw in the line of symmetry along the fold.

d Fold the rectangle in half a different way.
Unfold your shape.

e Using a dotted line, draw in the second line of symmetry.

The rectangle has two lines of symmetry:

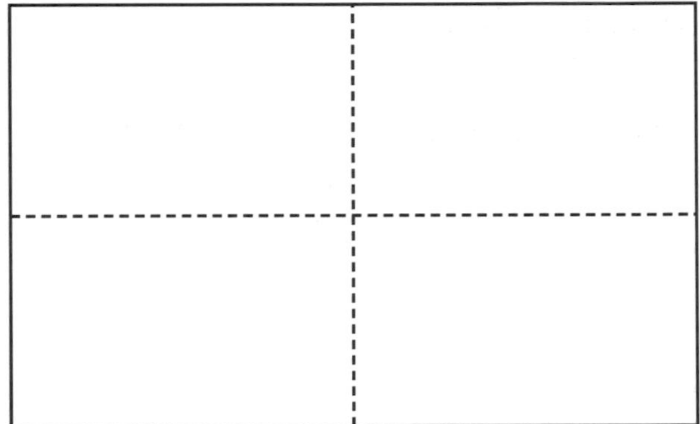

W 3

2 You will need copies of the shapes below.

a Cut out each shape.

b Find the lines of symmetry by folding your shape into two equal parts.

c Mark on the lines of symmetry with a dotted line.

(1)

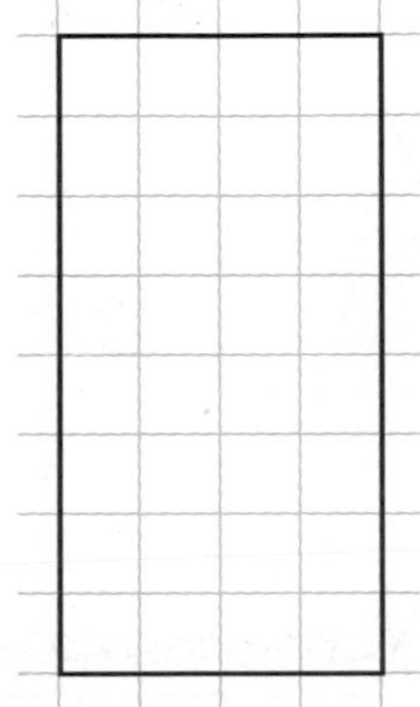

(2)

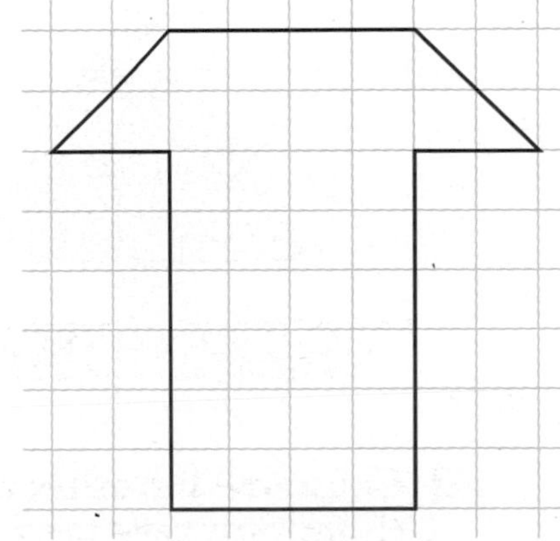

(3)

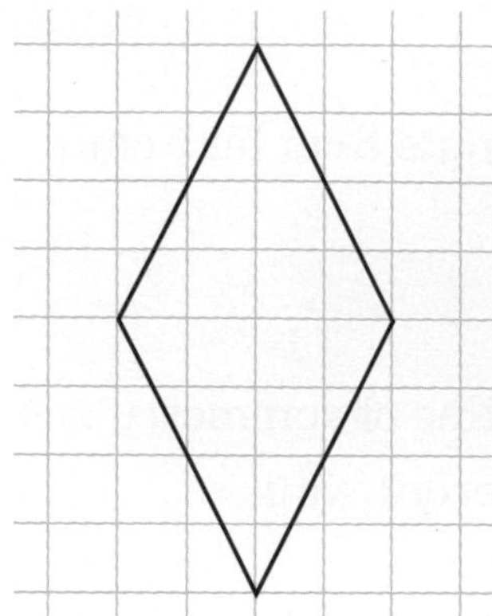

(7)

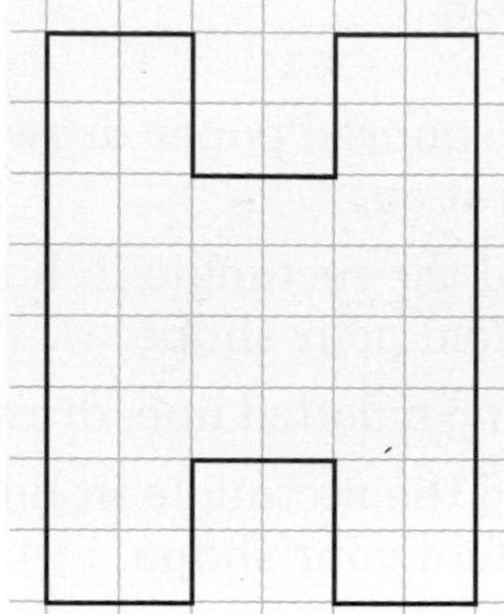

(4)

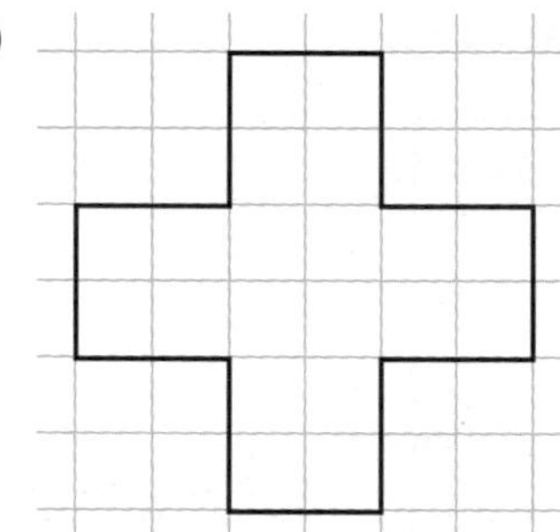

(8)

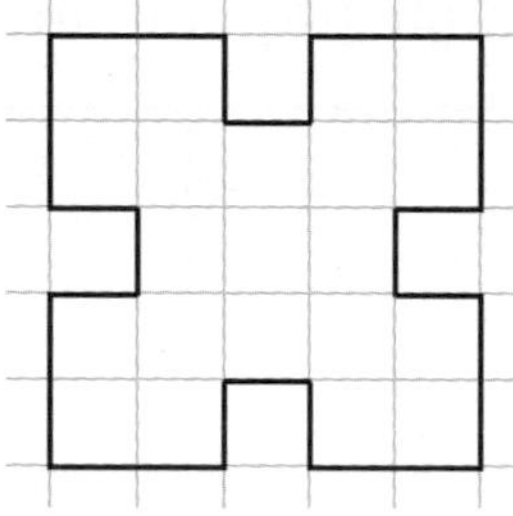

(5)

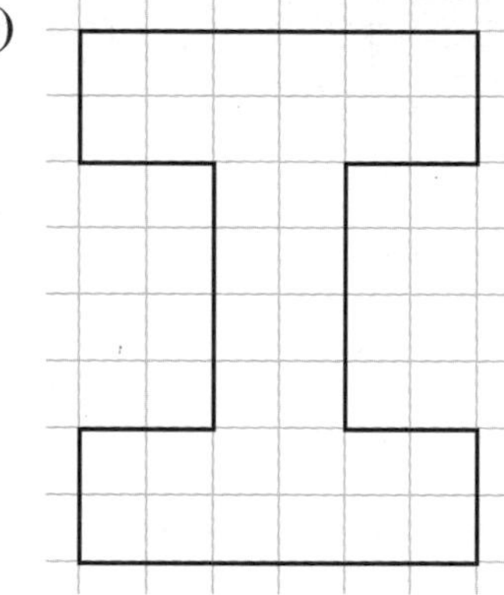

(9)

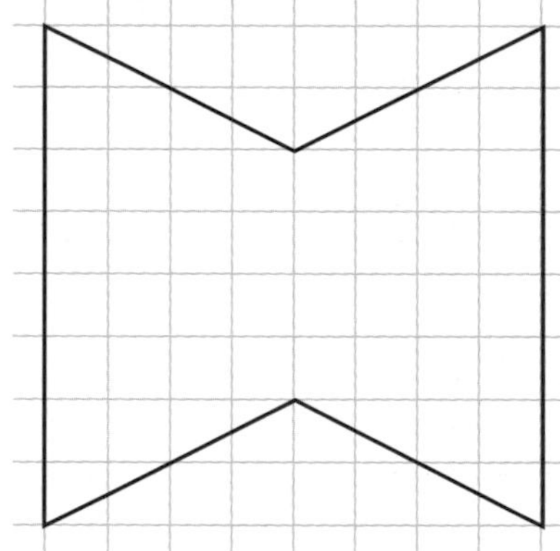

(6)

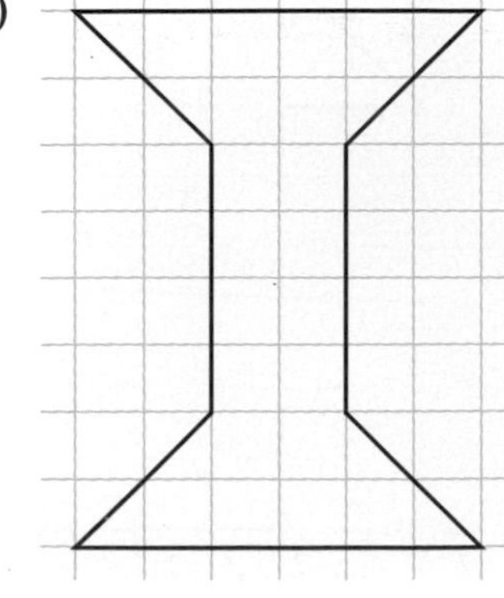

(10)

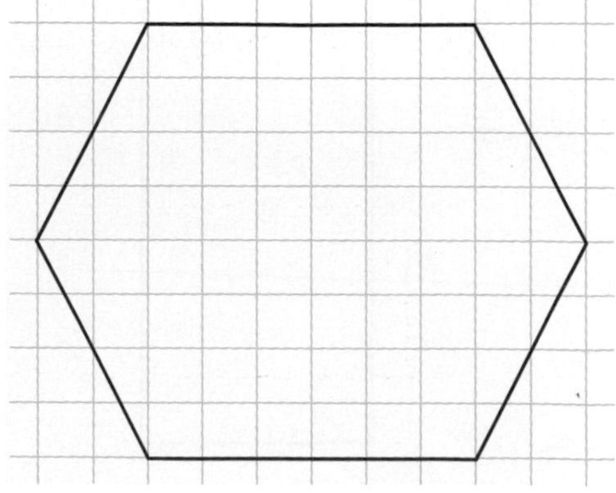

You can only see half of this shape:

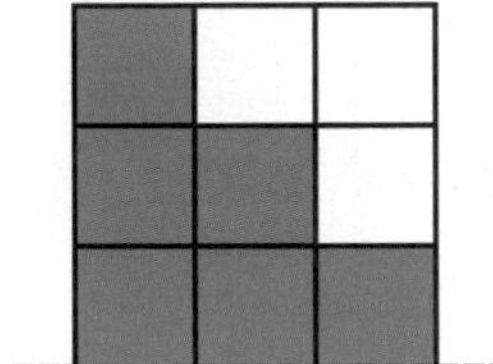

You can use a line of symmetry to complete the shape without cutting it out.
Use the squares to help you.

Three squares this way ⇒

so three squares this way ⇒

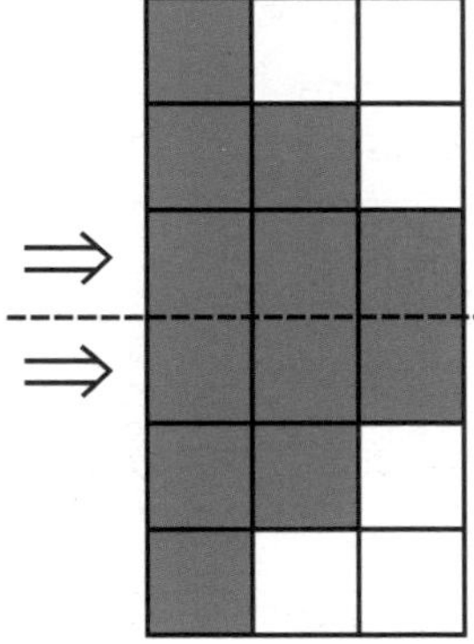

Exercise 2:6

4 **1** You need copies of these diagrams.
Use the line of symmetry to complete the shapes.
You can check them using tracing paper or a mirror.

a

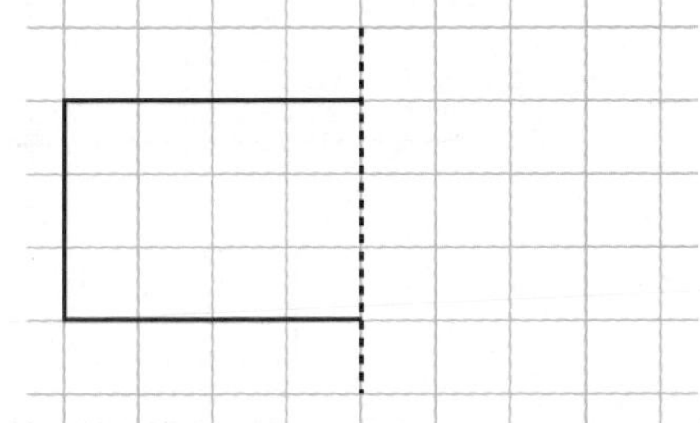

b

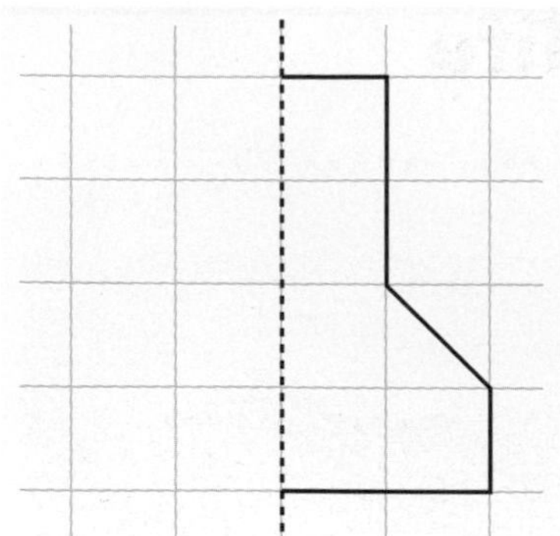

e

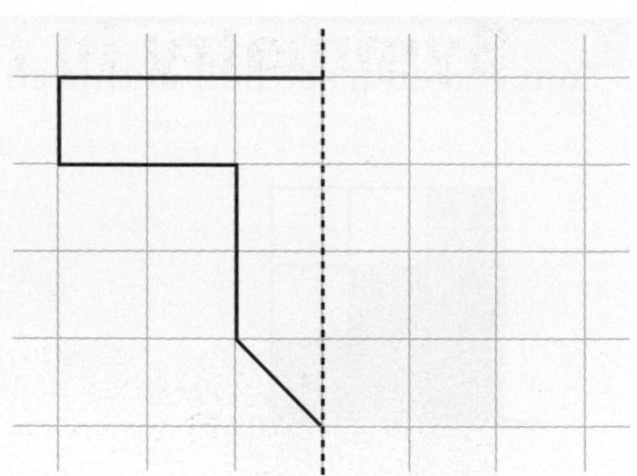

c

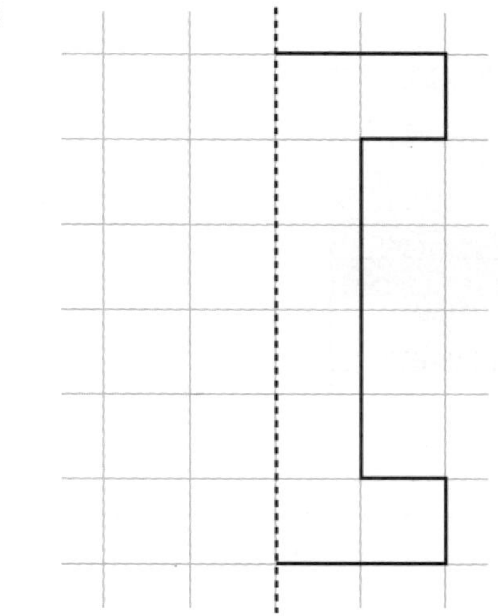

f

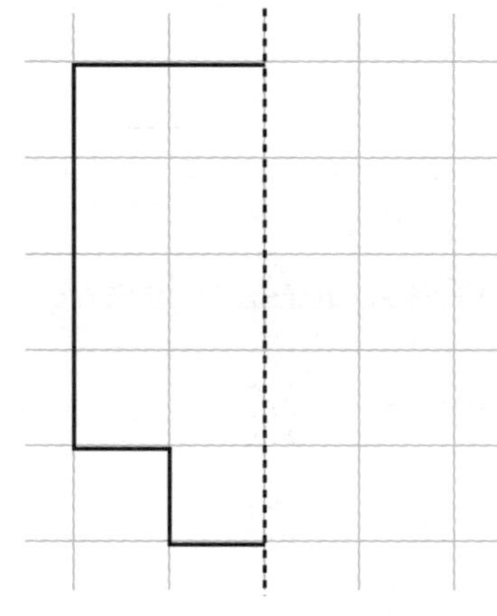

d

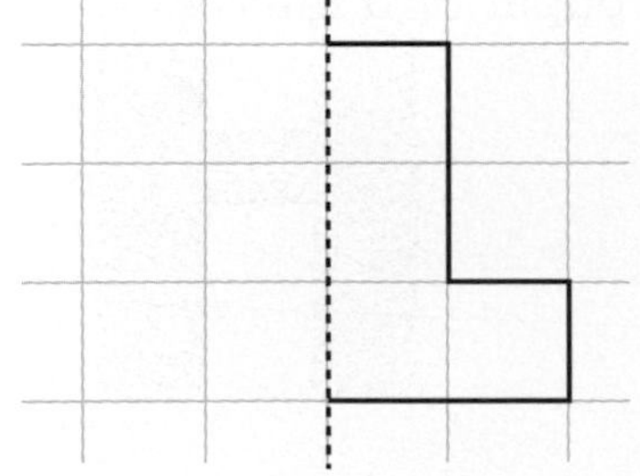

g

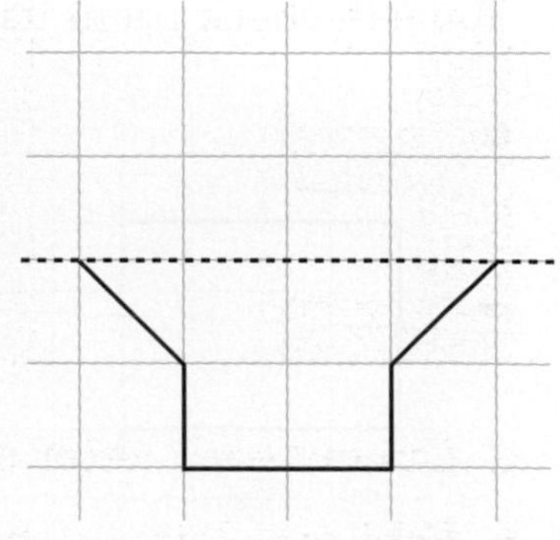

G 1, 2

2 Symmetry in real life

Here are some letters of the alphabet.

The letter  has one line of symmetry.

The letter  has two lines of symmetry.

The letter P has no lines of symmetry.

Exercise 2:7

W5 **1** On the worksheet there are some letters like these:

a Cut out each letter carefully.

b Fold each letter to find if it has a line of symmetry.

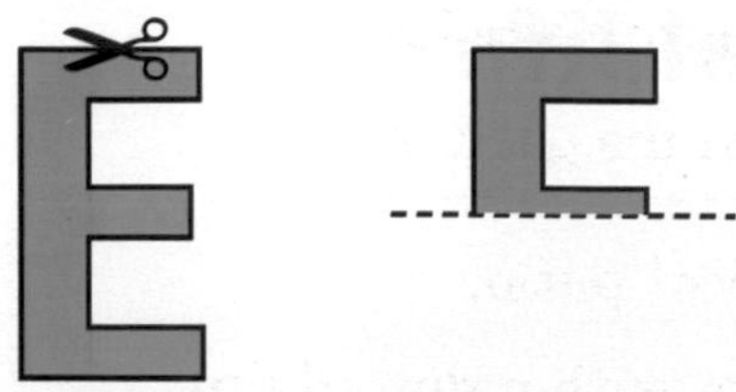

c Stick each letter in your book.

d Write by the side how many lines of symmetry the letter has.

E has one line of symmetry

 2

Exercise 2:8

 6

1 On the worksheet you will see some road signs.
Here is the first road sign:

a Trace each road sign carefully.
Peter says that you can test for symmetry by turning your tracing paper over to check if it has a line of symmetry.

b Do this for all the road signs on the worksheet.
Which signs have a line of symmetry?

W 7 **2** You will need copies of the diagrams below.

Mr Jones is building a patio in his garden.
He is using grey slabs.
He wants to have a symmetrical patio.

The line of symmetry is shown with a dotted line.
Complete the other side by counting the squares.

a

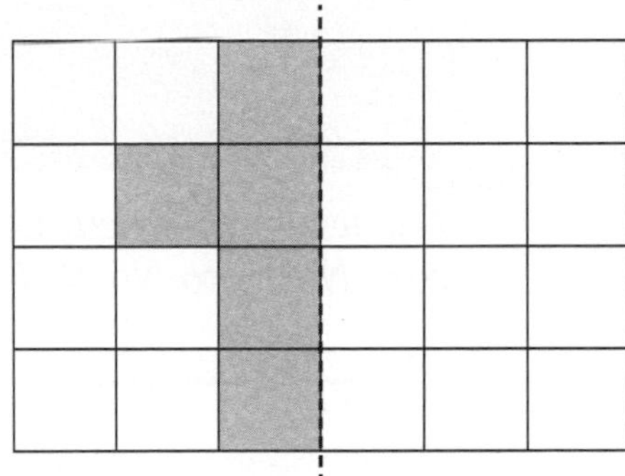

d

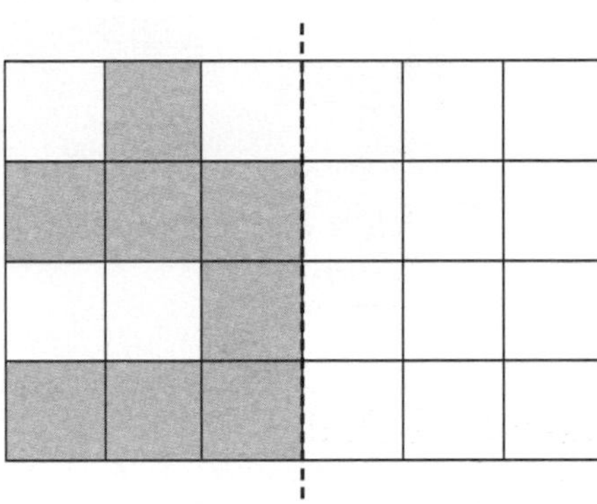

b

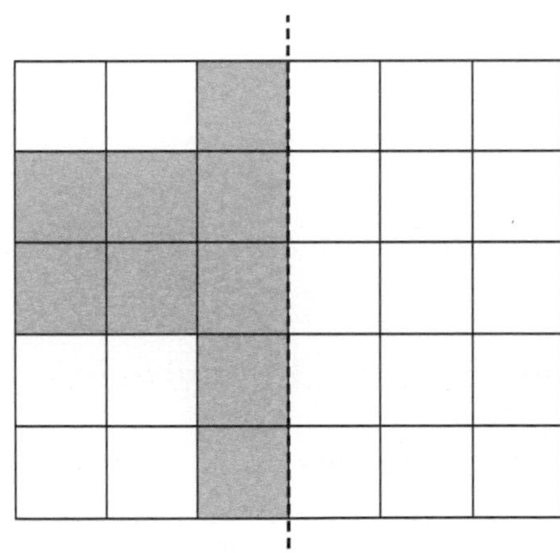

e

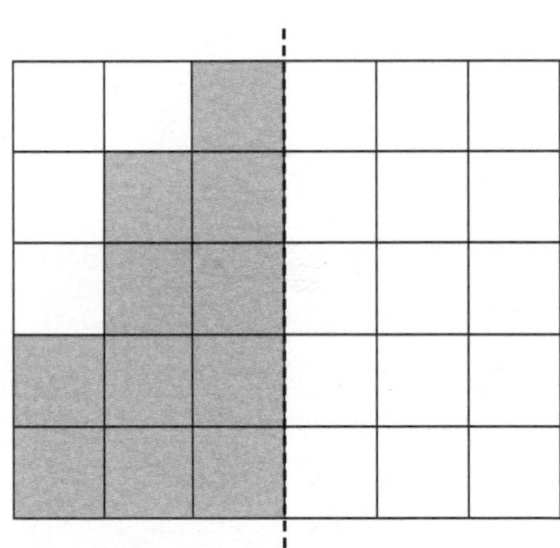

c

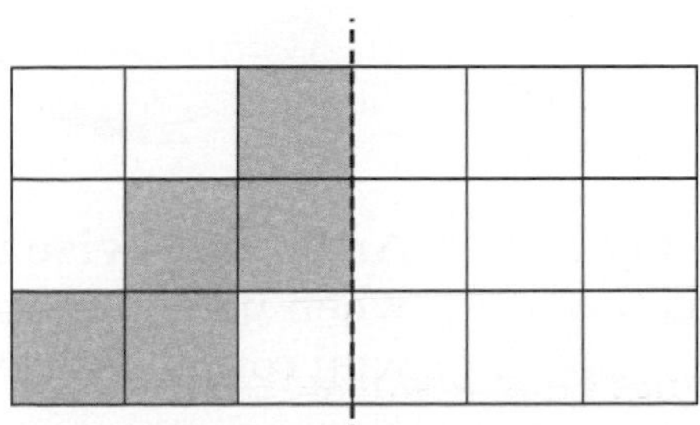

f

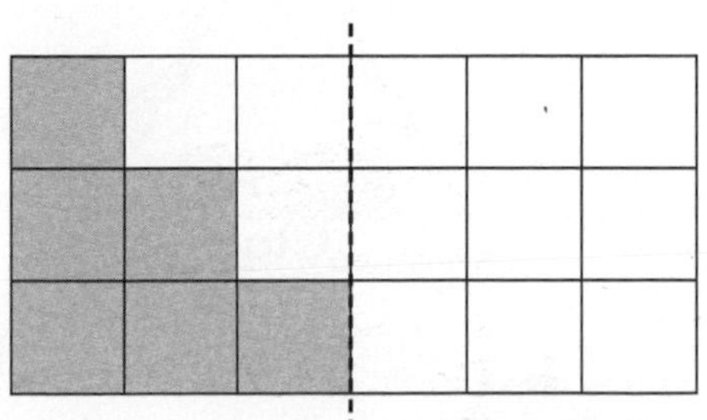

H 3

3 Turnings

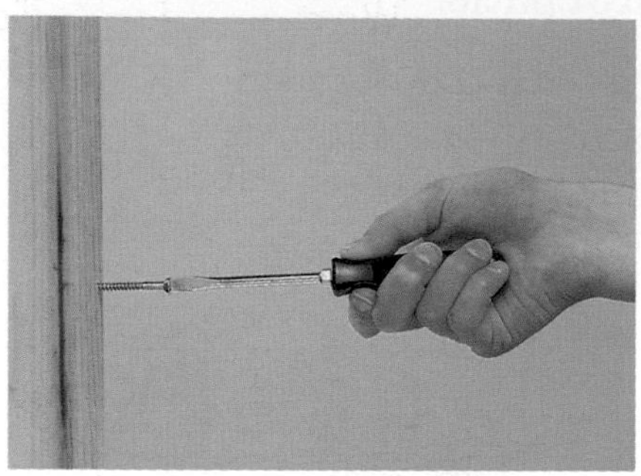

'Tighten the screw another half turn'

'The wind changed, turning from North to North East'

Turn

We use the word **turn** to describe something that moves round in a circle.

Start

1 full turn

Start

Half turn

Start

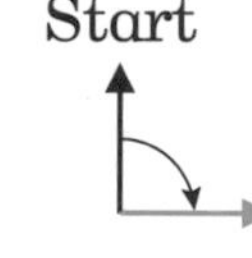

Quarter turn

Clockwise

We use **clockwise** and **anti-clockwise** to say which way to turn.

Clockwise is when you go the same way as the hands of the clock.

Anti-clockwise is when you go the other way round the clock.

Points of the compass

The diagram shows the **points of the compass**.

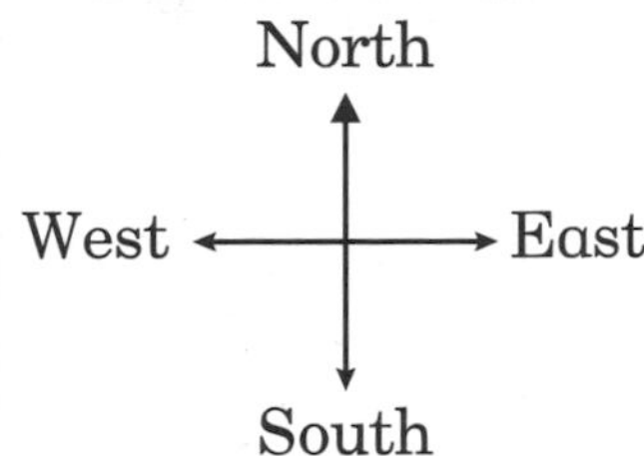

John is facing North

He turns a one-quarter turn clockwise.

He is now facing East.

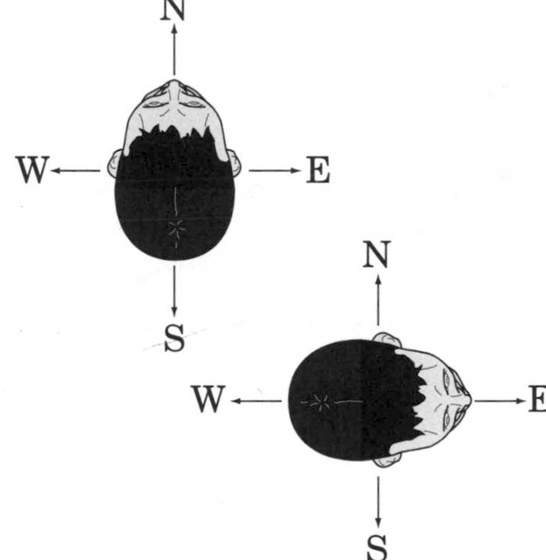

4

Exercise 2:9

1 **a** Draw Naomi facing North.
Naomi turns a half turn clockwise.

b Draw Naomi now.

c Which way is Naomi facing?

2 **a** Draw Carl facing South.
Carl turns a quarter turn clockwise.

b Draw Carl now.

c Which way is Carl facing?

3 **a** Draw Sophie facing West.
Sophie turns a three-quarter turn clockwise.

b Draw Sophie now.

c Which way is Sophie facing?

Sunita is facing East.

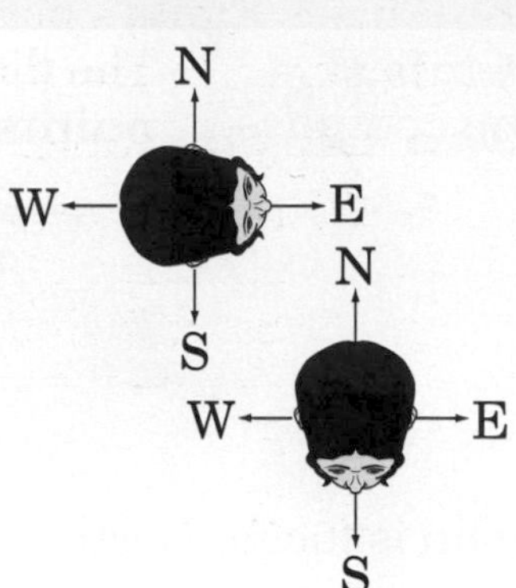

She turns a three-quarter turn anti-clockwise.

She is now facing South.

4 **a** Draw Megan facing West.
Megan turns a quarter turn anti-clockwise.
b Draw Megan now.
c Which way is Megan facing?

5 **a** Draw Matthew facing North.
Matthew turns a half turn anti-clockwise.
b Draw Matthew now.
c Which way is Matthew facing?

6 **a** Draw Jack facing South.
Jack turns a three-quarter turn anti-clockwise.
b Draw Jack now.
c Which way is Jack facing?

7 Copy and complete this table.

Start facing	Turn	Direction	End up facing
N	$\frac{1}{4}$	clockwise	
N	$\frac{1}{2}$	clockwise	
N	$\frac{3}{4}$	clockwise	
S	$\frac{1}{4}$	clockwise	
E	$\frac{3}{4}$	clockwise	
W	$\frac{1}{4}$	anti-clockwise	
S	$\frac{1}{2}$	anti-clockwise	
W	$\frac{3}{4}$	clockwise	
W	$\frac{3}{4}$	anti-clockwise	
S	$\frac{1}{4}$	anti-clockwise	

8 This is Emily's homework.
You need to mark her work.

a Copy the table.

b Write right or wrong for each question.

	Start facing	Turn	Direction	End up facing
(1)	S	$\frac{1}{4}$	clockwise	E
(2)	W	$\frac{1}{4}$	clockwise	N
(3)	S	$\frac{3}{4}$	clockwise	W
(4)	E	$\frac{1}{4}$	anti-clockwise	W
(5)	E	$\frac{1}{2}$	anti-clockwise	S
(6)	N	$\frac{1}{2}$	anti-clockwise	S

9 **a** Draw James facing South.
James turns a quarter turn anti-clockwise.

b Draw James now.

c Which way is James facing?

10 **a** Draw Gavin facing West.
Gavin turns a half turn clockwise.

b Draw Gavin now.

c Which way is Gavin facing?

11 **a** Draw Helen facing North.
Helen turns a three-quarter turn clockwise.

b Draw Helen now.

c Which way is Helen facing?

12 Copy and complete this table.

Start facing	Turn	Direction	End up facing
S	$\frac{3}{4}$	clockwise	
W	$\frac{1}{2}$	clockwise	
E	$\frac{1}{4}$	anti-clockwise	
N	$\frac{3}{4}$	anti-clockwise	
W	$\frac{1}{4}$	clockwise	
E	$\frac{3}{4}$	anti-clockwise	

W8 **1** You need copies of the shapes below.
Mark on **all** the lines of symmetry.

a

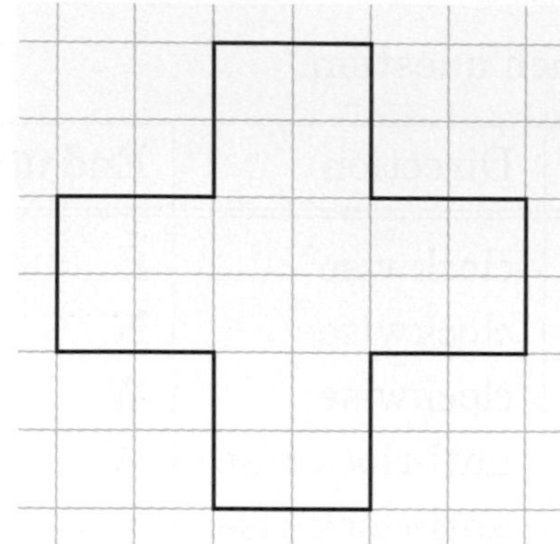

c

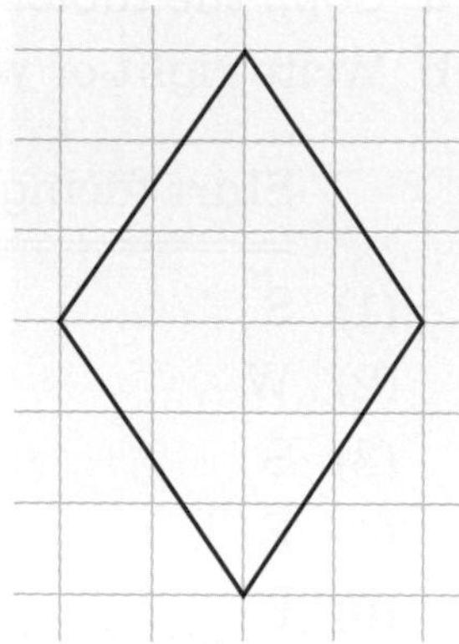

b

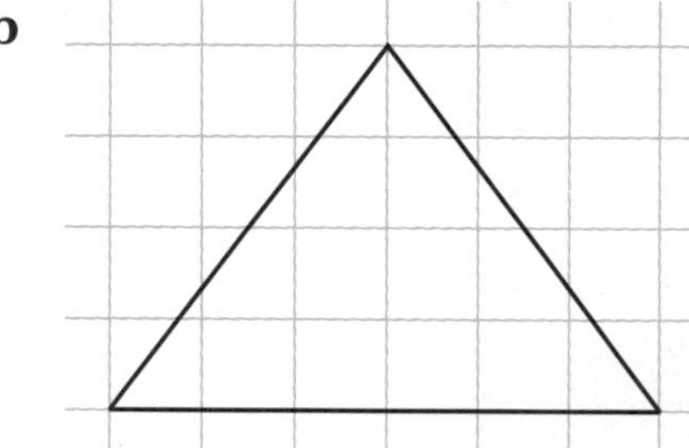

d

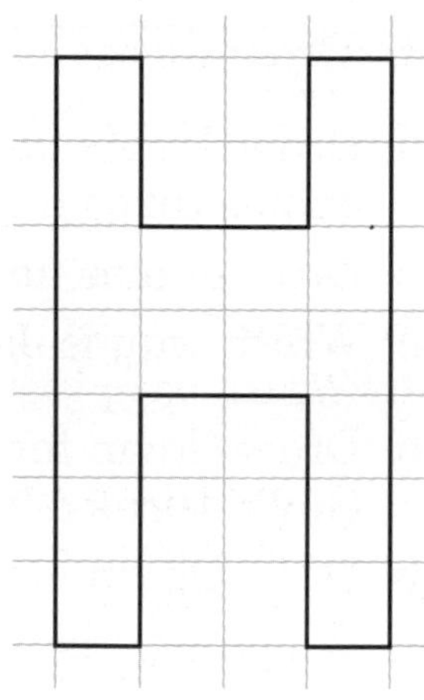

W9 **2** You need copies of the shapes below.
Draw their reflections in the line of symmetry.
You could use a mirror or tracing paper to help you.

a

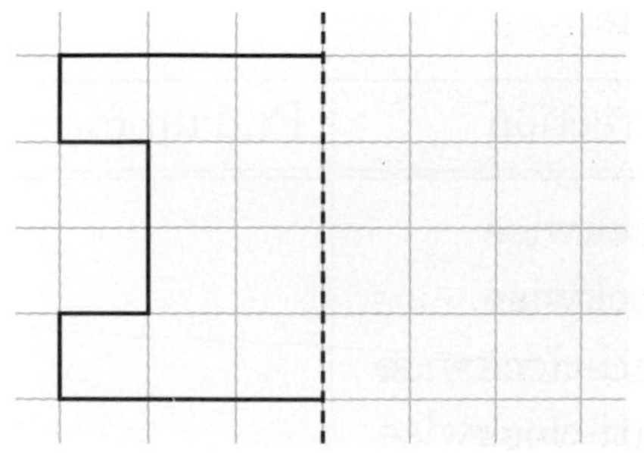

b

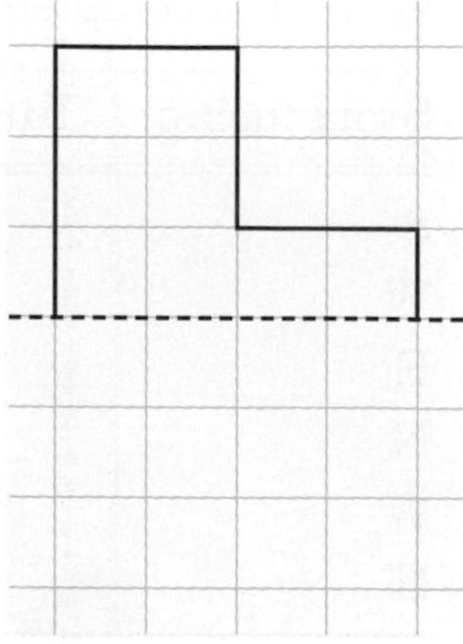

c

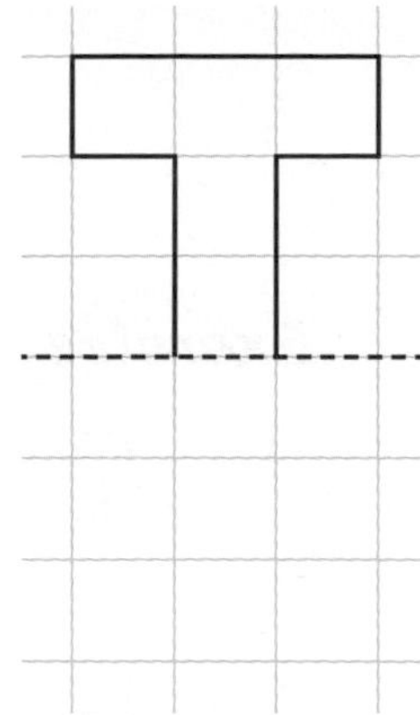

d

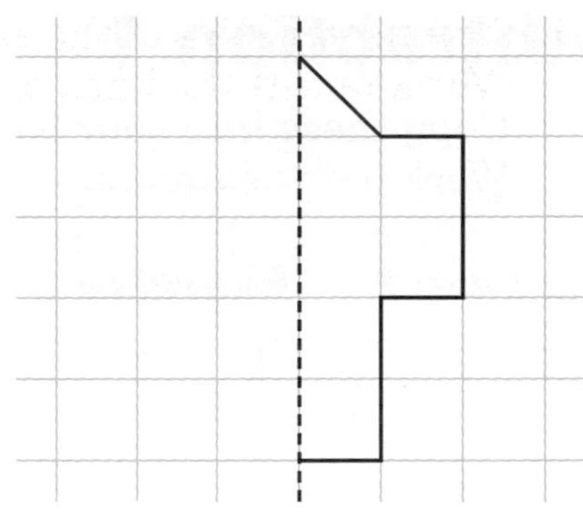

3 **a** Draw James facing South.
James turns a quarter turn anti-clockwise.
b Draw James now.
c Which way is James facing?

4 **a** Draw Megan facing West.
Megan turns a half turn clockwise.
b Draw Megan now.
c Which way is Megan facing?

5 **a** Draw Helen facing North.
Helen turns a three-quarter turn clockwise.
b Draw Helen now.
c Which way is Helen facing?

6 Copy and complete this table.

Start facing	Turn	Direction	End up facing
S	$\frac{3}{4}$	clockwise	
W	$\frac{1}{2}$	clockwise	
E	$\frac{1}{4}$	anti-clockwise	
N	$\frac{3}{4}$	anti-clockwise	
W	$\frac{1}{4}$	clockwise	
E	$\frac{3}{4}$	anti-clockwise	
S	$\frac{1}{2}$	anti-clockwise	
E	$\frac{1}{2}$	clockwise	
S	$\frac{1}{4}$	anti-clockwise	
N	$\frac{1}{4}$	clockwise	

Subtraction 1

Copy these into your book.
Work out the answers.

Exercise 1

1 7 − 5

2 8 − 3

3 9 − 4

4 10 − 2

5 10 − 7

6 15 − 2

7 18 − 4

8 19 − 7

9 16 − 5

10 20 − 10

Exercise 2

1 $\begin{array}{r} 17 \\ -\underline{\ 3} \\ \underline{\quad} \end{array}$

2 $\begin{array}{r} 15 \\ -\underline{\ 4} \\ \underline{\quad} \end{array}$

3 $\begin{array}{r} 19 \\ -\underline{\ 6} \\ \underline{\quad} \end{array}$

4 $\begin{array}{r} 24 \\ -\underline{\ 3} \\ \underline{\quad} \end{array}$

5 $\begin{array}{r} 28 \\ -\underline{\ 6} \\ \underline{\quad} \end{array}$

6 $\begin{array}{r} 29 \\ -\underline{\ 2} \\ \underline{\quad} \end{array}$

7 $\begin{array}{r} 25 \\ -\underline{12} \\ \underline{\quad} \end{array}$

8 $\begin{array}{r} 28 \\ -\underline{14} \\ \underline{\quad} \end{array}$

9 $\begin{array}{r} 29 \\ -\underline{17} \\ \underline{\quad} \end{array}$

10 $\begin{array}{r} 35 \\ -\underline{22} \\ \underline{\quad} \end{array}$

Exercise 3

1 13 − 5

2 14 − 7

3 12 − 6

4 15 − 8

5 16 − 9

6 22 − 5

7 23 − 8

8 25 − 6

9 27 − 19

10 30 − 19

Exercise 4

1 $\begin{array}{r} 10 \\ -\underline{\ 6} \\ \underline{\quad} \end{array}$

2 $\begin{array}{r} 12 \\ -\underline{\ 8} \\ \underline{\quad} \end{array}$

3 $\begin{array}{r} 14 \\ -\underline{\ 5} \\ \underline{\quad} \end{array}$

4 $\begin{array}{r} 17 \\ -\underline{\ 9} \\ \underline{\quad} \end{array}$

5 $\begin{array}{r} 23 \\ -\underline{\ 5} \\ \underline{\quad} \end{array}$

6 $\begin{array}{r} 24 \\ -\underline{\ 6} \\ \underline{\quad} \end{array}$

7 $\begin{array}{r} 23 \\ -\underline{\ 8} \\ \underline{\quad} \end{array}$

8 $\begin{array}{r} 25 \\ -\underline{\ 6} \\ \underline{\quad} \end{array}$

9 $\begin{array}{r} 27 \\ -\underline{\ 9} \\ \underline{\quad} \end{array}$

10 $\begin{array}{r} 33 \\ -\underline{\ 7} \\ \underline{\quad} \end{array}$

3 Number patterns

CORE

1 **Odds and evens**

2 **Patterns in number**

3 **Rules and robots**

QUESTIONS

HELP YOURSELF

Multiplication 1

This is a Chinese version of the number pattern known as 'Pascal's Triangle'. It appears in a manuscript dated 1303; Blaise Pascal was born in 1623.

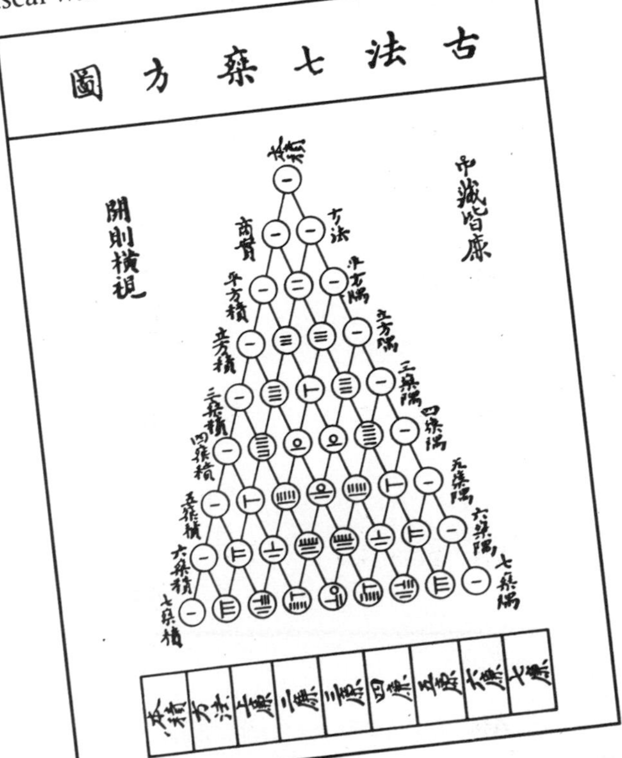

1 Odds and evens

Sometimes we cannot see any order or pattern.

Sometimes we can see order.

Counting numbers

The **counting numbers** are:

1 2 3 4 5 6 7 8 9 10 11 12 13 14 15 16 17 18 19 20 ...

Odd numbers

The **red** numbers have a pattern. They start the **odd numbers.**

1 3 5 7 9 11 13 15 17 19 ...

Even numbers

The blue numbers also have a pattern. They start the **even numbers.**

2 4 6 8 10 12 14 16 18 20 ...

Exercise 3:1

 1

1 **a** Copy this list of numbers into your book.

1 2 3 4 5 6 7 8 9 10 11 12 13 14 15
16 17 18 19 20

b Draw a red circle around the odd numbers.

c Draw a blue circle around the even numbers.

2 **a** Write the list of odd numbers in order starting with 1 and ending with 19.

b Write the list of even numbers in order starting with 2 and ending with 20.

3 **a** Which of these numbers are odd?

6 8 10 17 19 23 26

b Which of these numbers are even?

 1, 2

6 9 10 13 16 17 20

 1

4 Odd and even numbers can be bigger than 20.
Use a hundred square.
Colour the odd numbers in red.

 2

Colour the even numbers in blue.

When you use a **rule** you get a **number pattern.**

Each number in the pattern is called a **term**.

The **rule** for odd numbers is add 2.
The **rule** for even numbers is also add 2.

---

5 Write down the next two terms in each number pattern.

a 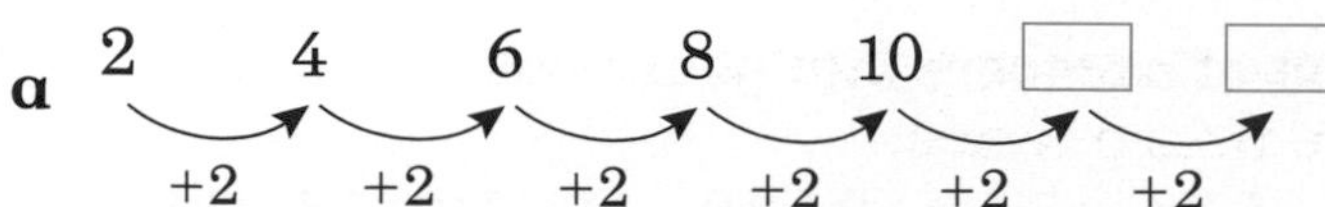

b 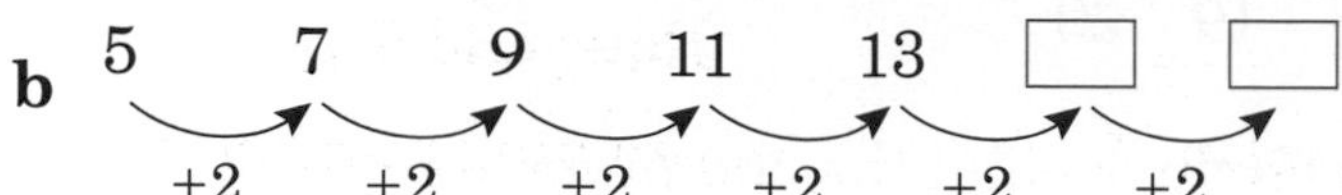

c 10 12 14 16 18 ☐ ☐

+2 +2 +2 +2 +2 +2

d 13 15 17 19 21 ☐ ☐

+2 +2 +2 +2 +2 +2

e 24 26 28 30 ☐ ☐

+2 +2 +2 +2 +2

f 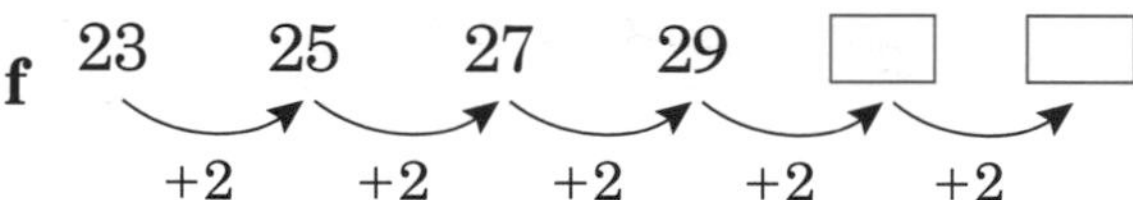

H 3

To make a **number pattern** you can use any rule and start with any number.

Examples **1** Starting number: (4)
Rule: add 3

Pattern: (4) 7 10 13 16 19

+3 +3 +3 +3 +3

This number pattern has 6 terms.

2 Starting number: (5)
Rule: add 4

Pattern: (5) 9 13 17 21

+4 +4 +4 +4

This number pattern has 5 terms.

Exercise 3:2

 4

1 Write down these number patterns.
Write down the first 6 terms.

a Starting number: 3
Rule: add 5

b Starting number: 6
Rule: add 2

c Starting number: 7
Rule: add 3

d Starting number: 4
Rule: add 5

e Starting number: 6
Rule: add 4

f Starting number: 5
Rule: add 7

Number patterns can go up or down.

Examples **1** Starting number: (20)
Rule: subtract 2

Pattern: (20) 18 16 14 12 10
(−2 between each term)

This number pattern has 6 terms.

2 Starting number: (24)
Rule: subtract 3

Pattern: (24) 21 18 15 12
(−3 between each term)

This number pattern has 5 terms.

 4

2 Write down these number patterns.
Write down the first 6 terms.

a Starting number: 16
Rule: subtract 1

b Starting number: 21
Rule: subtract 2

c Starting number: 35
Rule: subtract 5

d Starting number: 25
Rule: subtract 1

e Starting number: 32
Rule: subtract 3

f Starting number: 40
Rule: subtract 4

An **add rule** makes the number pattern go up.
A **subtract rule** makes the number pattern go down.

You can find the starting number and the rule from the number pattern.

Example

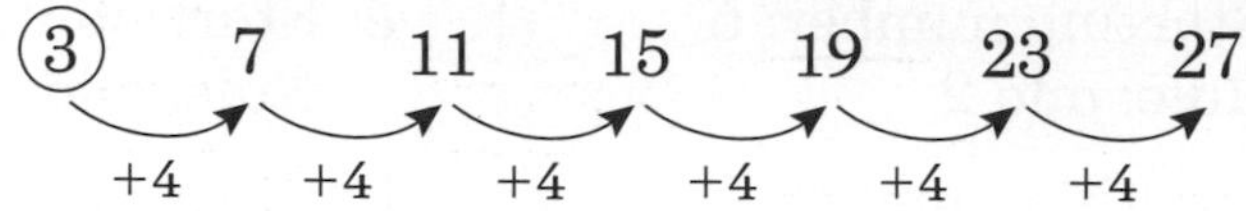

The starting number is ③

The number pattern is going up by 4 each time, so the rule is: add 4

This number pattern has 7 terms.

3 Find the starting number, the rule and the number of terms for each number pattern.

a

Starting number: ☐

Rule: ☐

Number of terms: ☐

b ㉒ 24 26 28

Starting number: ☐

Rule: ☐

Number of terms: ☐

H 5, 6

Multiples **Multiples** are special numbers.

Examples **1** A multiple of 2 can be made from 2s.

The multiples of 2 are:

2 4 6 8 10 …

2 A multiple of 3 can be made from 3s.

The multiples of 3 are:

3 6 9 12 15 …

You can find multiples in the times tables.

Exercise 3:3

7 **1** **a** Write these multiples of 2 in order, smallest first.

2 6 12 4 10 8

b Write these multiples of 3 in order, smallest first.

9 15 3 12 18 6

2 **a** Write down the numbers which are multiples of 2.

8 9 12 21 2 15

b Write down the numbers which are multiples of 3.

11 6 12 8 15 22

H 8 **c** Write down the numbers which are multiples of 5.

6 15 12 25 10 17

H 8 **d** Write down the numbers which are multiples of 10.

G 3, 4, 5 20 11 15 30 11 22 8

Factors A **factor** is a number that divides exactly into another number without a remainder.

Examples **1** Here are 10 counters made into different rectangles.

This rectangle has 2 rows of 5 counters.

This rectangle has 1 row of 10 counters.

The factors of 10 are 1, 2, 5 and 10.

2 3 counters can only be made into one rectangle.

The factors of 3 are 1 and 3.

Exercise 3:4

 9 **1** **a** Use counters to find the factors of 6.

b Use counters to find the factors of 5.

c Use counters to find the factors of 15.

G 6, 7, 8, 9 **d** Use counters to find the factors of 9.

Prime numbers

A **prime number** has only two factors, itself and 1.
1 is not a prime number.

Prime numbers can only be made into one rectangle with counters.

7 has only 1 rectangle.
7 is a prime number.

Exercise 3:5

1 Find out which of these numbers are prime numbers.
You can use counters to help you.

a 5

b 4

c 11

d 14

e 16

f 17

2 Write down these prime numbers in order, smallest first.

13 2 5 23 17 7 11 19 3 29

3 There is only one even number that is prime.
Write down this number.

2 Patterns in number

Some number patterns have special names. The names describe the shapes that they make.

Square numbers

You can make **square numbers** by putting counters into squares.

4 is a square number

9 is a square number

16 is a square number

Exercise 3:6

 9

1 Find out which of these numbers are square numbers.
You can use counters to help you.

a 25 **d** 36

b 49 **e** 28

c 12 **f** 18

You can find the square numbers by multiplying.
Multiply two numbers that are the same.

Example $1 \times 1 = \boxed{1}$ **1** is a square number

$2 \times 2 = \boxed{4}$ **4** is a square number

$3 \times 3 = \boxed{9}$ **9** is a square number

2 Copy and complete.

a $4 \times 4 = \square$ $\square$ is a square number.

b $5 \times 5 = \square$ $\square$ is a square number.

c $6 \times 6 = \square$ $\square$ is a square number.

Triangle numbers

You can make **triangle numbers** by putting counters into triangles.

3 is a triangle number

6 is a triangle number

10 is a triangle number

Exercise 3:7

 9

1 Find out which of these numbers are triangle numbers.
You can use counters to help you.

a 5

b 10

c 17

d 21

You can find the triangle numbers by adding.

Example	$\boxed{1}$	**1** is a triangle number
	$1 + 2 = \boxed{3}$	**3** is a triangle number
	$1 + 2 + 3 = \boxed{6}$	**6** is a triangle number

---

2 Copy and complete.

a $1 + 2 + 3 + 4 = \square$ $\square$ is a triangle number.

b $1 + 2 + 3 + 4 + 5 = \square$ $\square$ is a triangle number.

 10

c $1 + 2 + 3 + 4 + 5 + 6 = \square$ $\square$ is a triangle number.

Patterns from numbers

You will need the circle worksheets.

W 2

1 Odd and even pattern
Draw a blue dot on each even number on the first circle.
Join the dots to make a shape.

Draw a red dot on each odd number.
Join the dots to make a shape.

Your shapes should look like this.

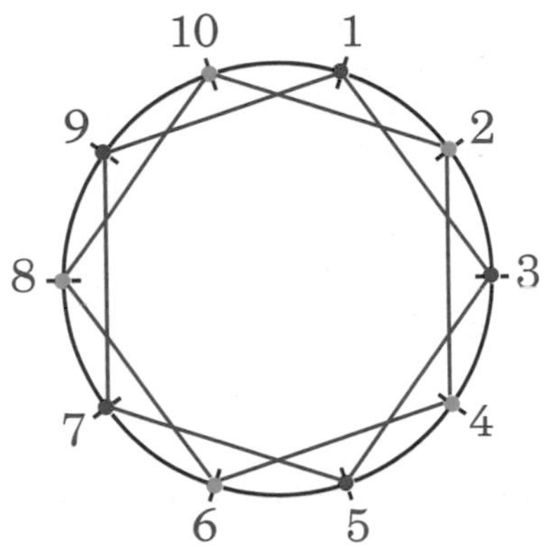

W 2

2 Multiple patterns

a Draw a dot on each multiple of 2 on the second circle.
Join the dots to make a shape.
What shape does it make?

b Join the multiples of 3 on the third circle.
What shape does it make?

c Join the multiples of 4 on the fourth circle.
What shape does it make?

W 3

3 Factor patterns

a Join the factors of 6 on the first circle.

b Join the factors of 8 on the second circle.

c Join the factors of 14 on the third circle.

d Join the factors of 15 on the fourth circle.

e Write down how many sides each shape has.

W 4

4 More factor patterns
Draw factor shapes for

5 9 10 11 12

Write down how many sides each shape has.

5 Make a poster of your shapes.
Write some sentences about what you have found out.

3 Rules and robots

Imagine we have a robot to help us with our patterns.

This robot adds 5 to any number.

Example What answers will this robot give?

We only need to draw the robot's screen.

5 → +3 → 8

Exercise 3:8

Draw the screen for each robot.
Write down the answer for each robot.

1 **a** 1 → +2 → ?

b 3 → +2 → ?

c 5 → +2 → ?

2 **a** 2 → +7 → ?

b 4 → +7 → ?

c 5 → +7 → ?

3 **a** 3 → +10 → ?

b 4 → +10 → ?

c 8 → +10 → ?

4 **a** 3 → −1 → ?

b 5 → −1 → ?

c 10 → −1 → ?

W 5

G 10, 11, 12, 13, 14, 15

One day, the robot's screen does not work.
You can work out what the screen should show.

3 → ? → 8

5 → ? → 10

The screens should show +5

Exercise 3:9

Write down the rule that belongs to each screen.

1 1 → ? → 5

3 → ? → 7

2 2 → ? → 8

6 → ? → 12

3 5 → ? → 12

8 → ? → 15

4 5 → ? → 3

8 → ? → 6

5 20 → ? → 16

13 → ? → 9

6 30 → ? → 10

25 → ? → 5

W 6

You can use two robots.

4 → ×3 → 12 → +1 → 13

2 → ×3 → 6 → +1 → 7

Exercise 3:10

1 You need copies of these robot screens.
Complete the screens.

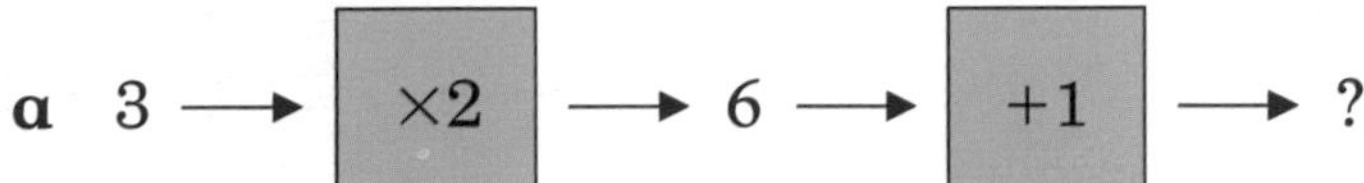

b 4 → ×2 → ? → +1 → ?

c 5 → ×2 → ? → +1 → ?

2 You need copies of these robot screens.
Complete the screens.

a 2 → −1 → 1 → ×3 → ?

b 3 → −1 → ? → ×3 → ?

c 6 → −1 → ? → ×3 → ?

3 You need copies of these robot screens.
Complete the screens.

a 2 → +4 → ? → ×2 → ?

b 4 → +4 → ? → ×2 → ?

c 5 → +4 → ? → ×2 → ?

1 **a** Which of these numbers are odd?

6 9 10 13 16 17 20

b Which of these numbers are even?

3 6 8 10 14 17 19

2 **a** Which of these numbers are odd?

15 17 22 26 31 45 66 77 82 91

b Which of these numbers are even?

22 25 46 57 68 76 79 80 93 100

3 Write down these number patterns.
Write down the first 6 terms.

a Starting number: (2)
Rule: add 3

b Starting number: (10)
Rule: add 5

c Starting number: (6)
Rule: add 1

d Starting number: (3)
Rule: add 2

4 Write down these number patterns.
Write down the first 6 terms.

a Starting number: (20)
Rule: subtract 2

b Starting number: (40)
Rule: subtract 5

c Starting number: (60)
Rule: subtract 10

d Starting number: (13)
Rule: subtract 2

5 Look at this list of numbers:

10 11 12 13 14 15 16 17 18 19 20

a Which of these numbers are multiples of 2?

b Which of these numbers are multiples of 3?

c Which of these numbers are multiples of 5?

d Which of these numbers are multiples of 10?

6 Write down the factors of 12.

7 Write down the factors of 20.

8 Look at this list of numbers:

1 5 9 13 17 21 25

a Which of these numbers are prime numbers?

b Which of these numbers are square numbers?

c Which of these numbers are triangle numbers?

9 Copy the screen for each robot.
Write down the answer for each robot.

a 4 → [+3] → ?

b 5 → [+3] → ?

c 3 → [×10] → ?

d 5 → [×10] → ?

10 This robot has broken down.
Write down the rule that should be on each screen.

12 → [?] → 8

7 → [?] → 3

11 Copy the diagrams for each robot.
Write down the answers.

a 1 → [+4] → [×3] → ?

b 2 → [+3] → [×5] → ?

c 4 → [−2] → [×2] → ?

Multiplication 1

Copy these into your book.
Work out the answers.

Exercise 1	**Exercise 2**	**Exercise 3**	**Exercise 4**
1 $1\times 2=$	**1** $8\times 5=$	**1** $8\times 2=$	**1** $7\times 5=$
2 $2\times 5=$	**2** $6\times 10=$	**2** $6\times 5=$	**2** $9\times 2=$
3 $4\times 2=$	**3** $1\times 5=$	**3** $6\times 2=$	**3** $4\times 2=$
4 $4\times 5=$	**4** $4\times 10=$	**4** $7\times 5=$	**4** $3\times 10=$
5 $2\times 10=$	**5** $7\times 2=$	**5** $8\times 10=$	**5** $4\times 10=$
6 $3\times 2=$	**6** $2\times 2=$	**6** $1\times 2=$	**6** $5\times 5=$
7 $9\times 5=$	**7** $4\times 5=$	**7** $1\times 10=$	**7** $2\times 2=$
8 $7\times 10=$	**8** $5\times 2=$	**8** $4\times 5=$	**8** $10\times 2=$
9 $10\times 2=$	**9** $3\times 5=$	**9** $3\times 10=$	**9** $7\times 2=$
10 $5\times 5=$	**10** $6\times 2=$	**10** $10\times 5=$	**10** $6\times 2=$
11 $9\times 2=$	**11** $9\times 10=$	**11** $8\times 5=$	**11** $6\times 10=$
12 $6\times 5=$	**12** $1\times 2=$	**12** $3\times 5=$	**12** $7\times 10=$
13 $5\times 10=$	**13** $10\times 2=$	**13** $4\times 2=$	**13** $2\times 5=$
14 $8\times 5=$	**14** $2\times 10=$	**14** $3\times 2=$	**14** $8\times 5=$
15 $5\times 2=$	**15** $10\times 10=$	**15** $1\times 5=$	**15** $10\times 10=$
16 $8\times 2=$	**16** $9\times 2=$	**16** $9\times 5=$	**16** $1\times 2=$

4 Arithmetic and the calculator

CORE

1 **Rounding up and rounding down**

2 **Mental arithmetic**

3 **Estimation**

QUESTIONS

HELP YOURSELF

Division 1

John Napier was born in Scotland in 1550.
One of the calculating methods he invented is known as Napier's Bones or Napier's Rods.

0	1	2	3	4	5	6	7	8	9	Index
0/0	0/1	0/2	0/3	0/4	0/5	0/6	0/7	0/8	0/9	1
0/0	0/2	0/4	0/6	0/8	1/0	1/2	1/4	1/6	1/8	2
0/0	0/3	0/6	0/9	1/2	1/5	1/8	2/1	2/4	2/7	3
0/0	0/4	0/8	1/2	1/6	2/0	2/4	2/8	3/2	3/6	4
0/0	0/5	1/0	1/5	2/0	2/5	3/0	3/5	4/0	4/5	5
0/0	0/6	1/2	1/8	2/4	3/0	3/6	4/2	4/8	5/4	6
0/0	0/7	1/4	2/1	2/8	3/5	4/2	4/9	5/6	6/3	7
0/0	0/8	1/6	2/4	3/2	4/0	4/8	5/6	6/4	7/2	8
0/0	0/9	1/8	2/7	3/6	4/5	5/4	6/3	7/2	8/1	9

You can make a set of 'bones' on paper or card like this.
Can you find out how they work?

1 Rounding up and rounding down

50 people attend a school reunion

This is a newspaper headline.
The exact number of people may not be 50.
It has been rounded to the nearest ten.
We don't know the exact number, it could be 49 or 52 or any other number that rounds to 50.
We can round numbers to the nearest 10.

Look at this number line.

It shows the tens.

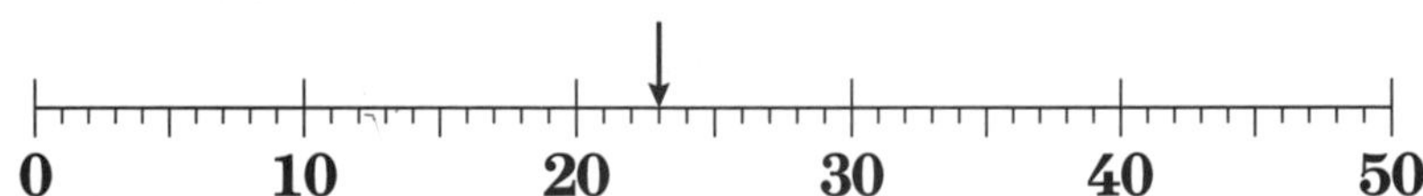

23 has been marked with an arrow.

Exercise 4:1

W 1

1 You need a 0 to 50 number line.
Mark these numbers on a number line with an arrow.

a 12
b 37
c 29
d 48
e 43
f 8
g 17
h 2

Example Here is a number line.
It shows the tens.
23 has been marked with an arrow.

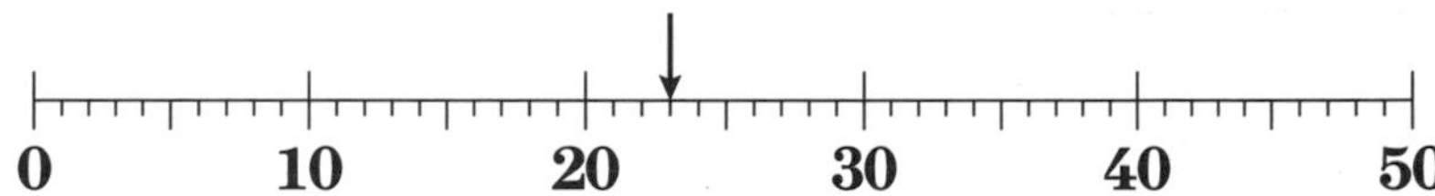

23 is between 20 and 30. Draw a ring around the nearest tens.

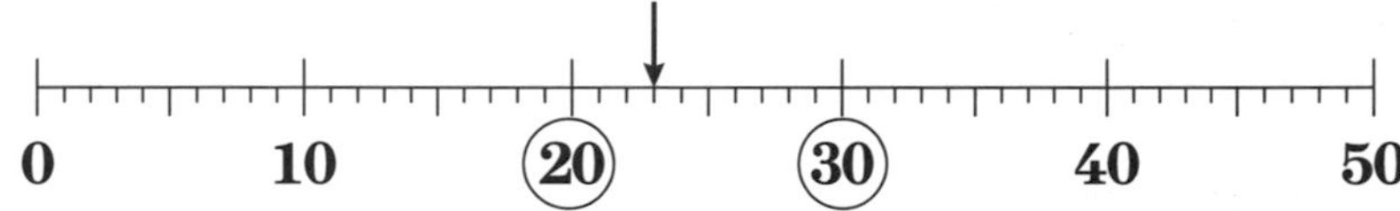

23 is closer to 20. 23 is rounded **down** to 20.

W 1

2 You need a 0 to 50 number line for each part of this question.
Mark these numbers on a number line.
Draw rings around the nearest tens.
Round them to the nearest 10.

a 11
b 32
c 44
d 13
e 21
f 42
g 14
h 33

Example Here is a number line.
It shows the tens.
47 has been marked with an arrow.

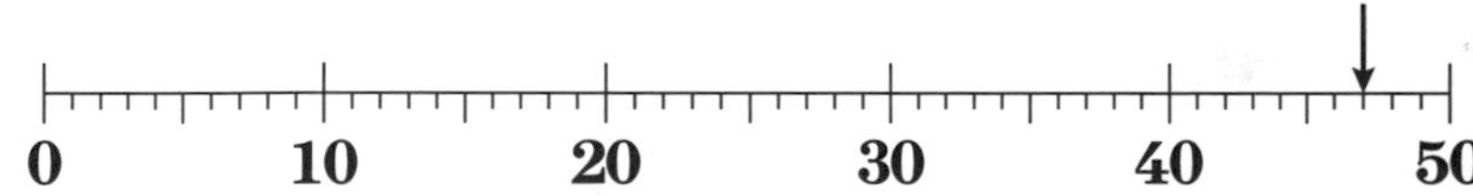

47 is between 40 and 50. Draw a ring around the nearest tens.

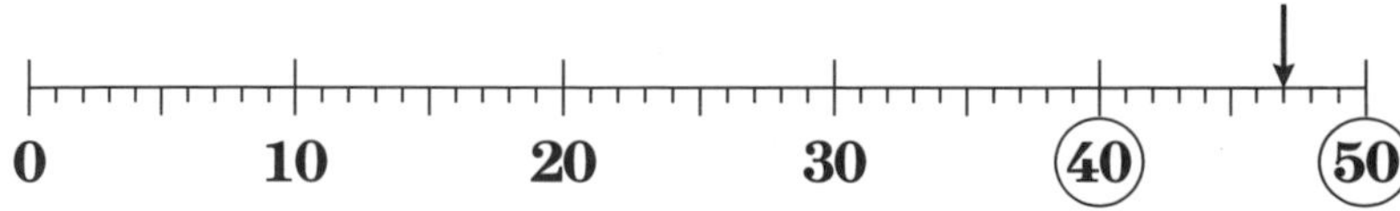

47 is closer to 50. 47 is rounded **up** to 50.

W 1 **3** You need a 0 to 50 number line for each part of this question.
Mark these numbers on a number line.
Draw rings around the nearest tens.
Round them to the nearest 10.

a 18 **e** 28
b 36 **f** 46
c 49 **g** 19
d 7 **h** 37

W 1 **4** Mark these numbers on a number line.
Draw rings around the nearest tens.
Round them to the nearest 10.
Some will be rounded up and some will be rounded down.

a 4 **e** 24
b 29 **f** 38
c 41 **g** 19
G 1 **d** 9 **h** 27

Here is a number line.
25 has been marked with an arrow.
25 is half-way between 20 and 30.
Numbers that are half-way are always rounded **up**.

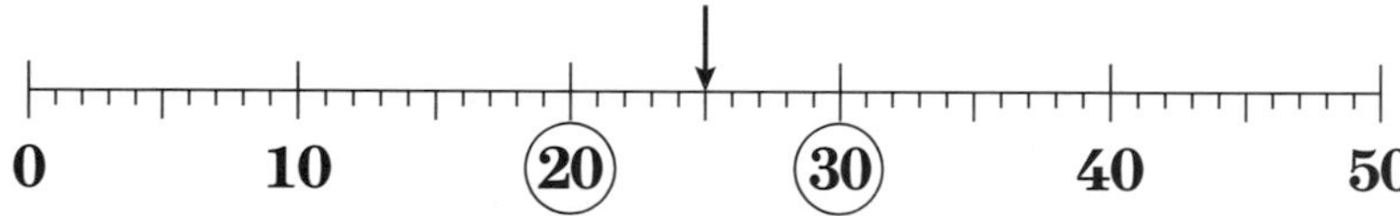

25 is rounded **up** to 30.

---

5 Round these numbers to the nearest 10.

a 25
b 35
c 15
d 5
e 45

You can round numbers to the nearest 100.

Example Look at this number line.
It shows the hundreds.
240 has been marked with an arrow.

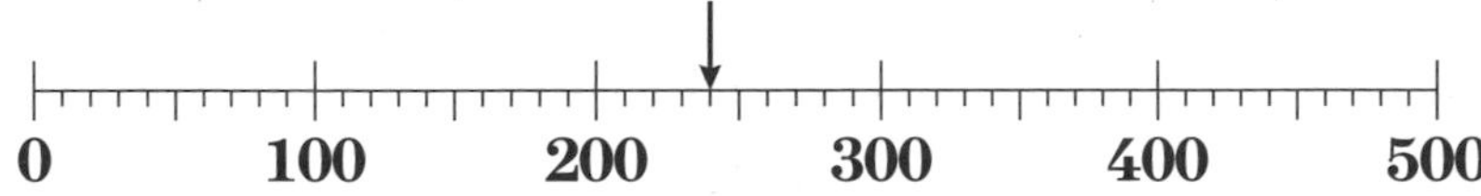

240 is between 200 and 300. Draw a ring around the nearest hundreds.

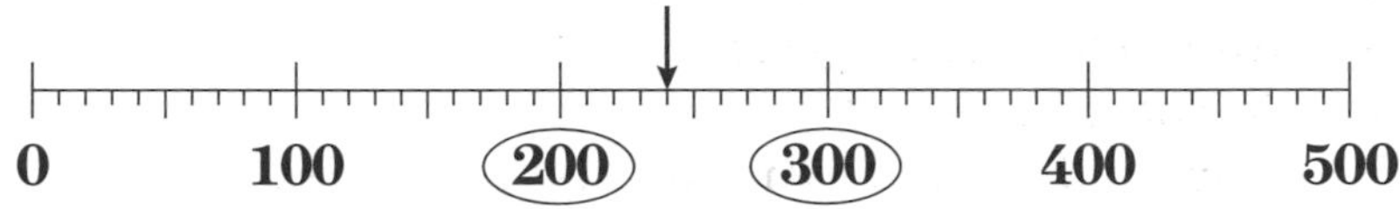

240 is closer to 200. 240 is rounded **down** to 200.

Exercise 4:2

2 **1** You need a hundreds number line for each part of this question.
Mark these numbers on a number line with an arrow.
Draw a ring around the nearest hundreds.
Round them to the nearest hundred.

a 120 **b** 340 **c** 230 **d** 410

Example Here is a number line.
It shows the hundreds.
170 has been marked with an arrow.

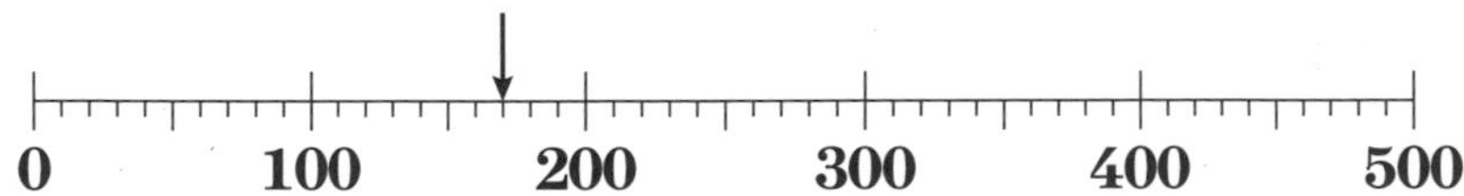

170 is between 100 and 200. Draw a ring around the nearest hundreds.

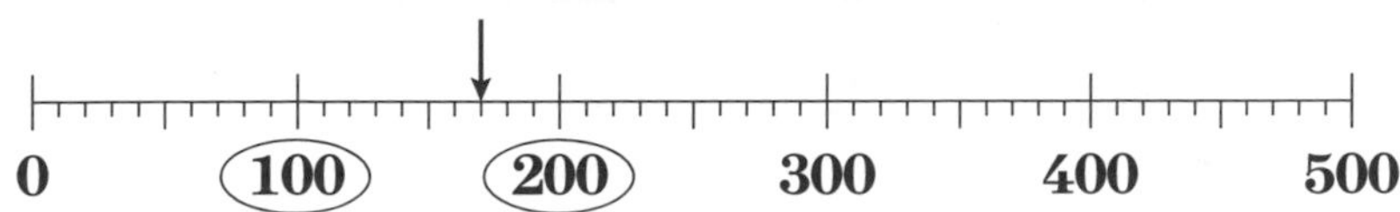

170 is closer to 200. 170 is rounded **up** to 200.

W 2

2 You need a hundreds number line for each part of this question.
Mark these numbers on a number line.
Draw rings around the nearest hundreds.
Round them to the nearest 100.

a 260
b 370
c 190
d 480

Here is a number line.
250 has been marked with an arrow.
250 is half-way between 200 and 300.
Numbers that are half-way are always rounded **up**.

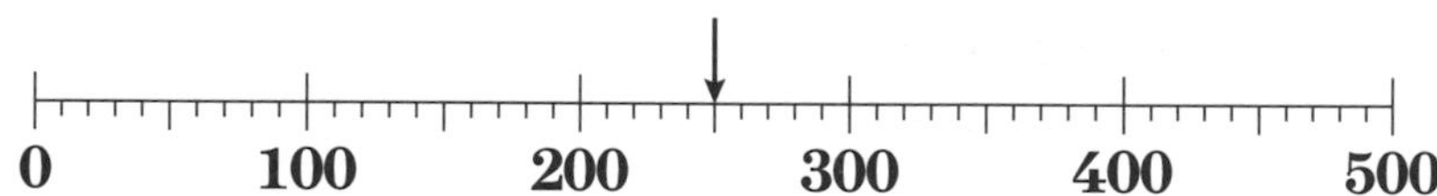

250 is rounded **up** to 300.

---

3 Round these numbers to the nearest 100.

a 250
b 350
c 150
d 50
e 450

Look at this number line.
Not all the numbers are on the line.
27 has been marked with an arrow.

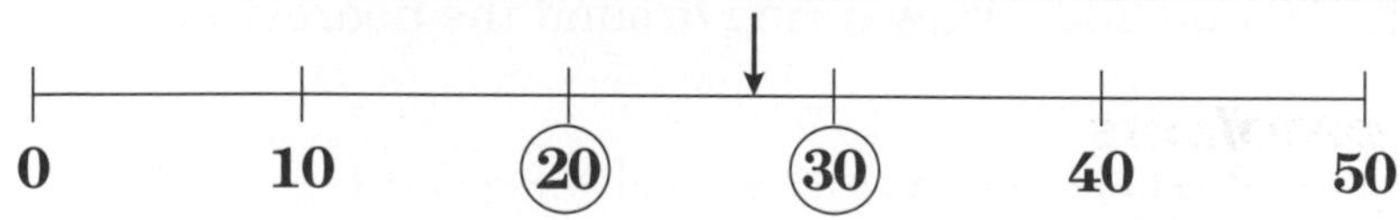

27 is between 20 and 30.
This is called an **estimate**.

Exercise 4:3

 3 **1** You need a 0 to 50 number line.
Mark these numbers on a number line.

a 24	**e** 33
b 39	**f** 18
c 41	**g** 13
d 9	**h** 27

Look at this number line.
Not all the numbers are on the line.
330 has been marked with an arrow.

330 is between 300 and 400.
This is called an **estimate**.

 4 **2** You need a hundreds number line.
Mark these numbers on a number line.

a 240	**e** 330
b 390	**f** 180
c 410	**g** 135
d 90	**h** 275

Rounding in problems

You can use rounding to help you work things out without a calculator.
These examples are all in a PE department where the different balls need sorting out.

This is a tennis ball. They come in boxes of 5.

 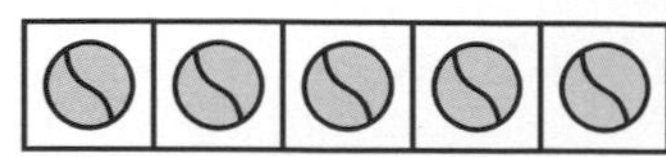

The PE department has
12 tennis balls:
they need 3 boxes to put
the balls away.

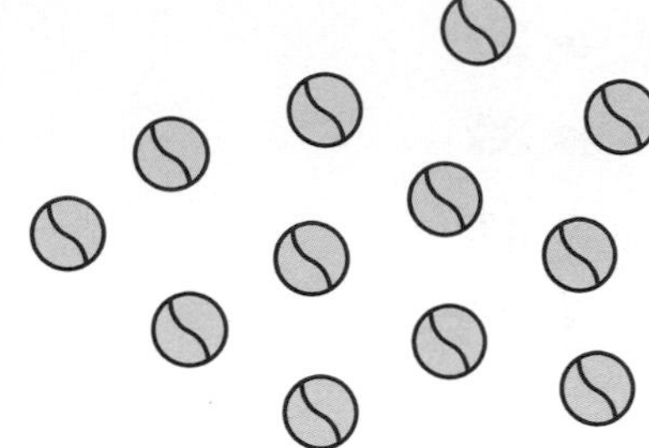

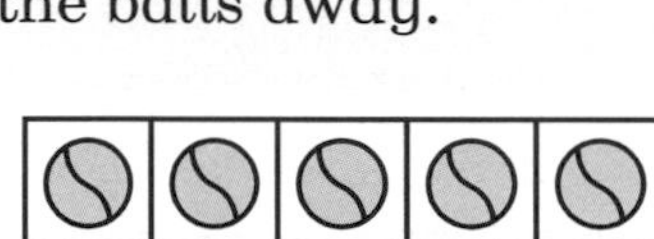

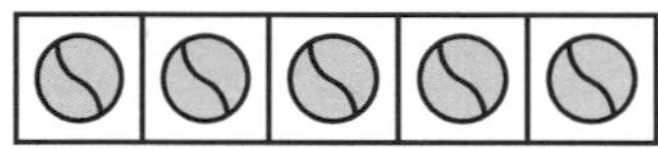

Exercise 4:4

1 How many boxes do they need for these tennis balls?

a 9 **d** 21

b 17 **e** 10

c 4 **f** 14

This is a squash ball. They come in boxes of 3.

 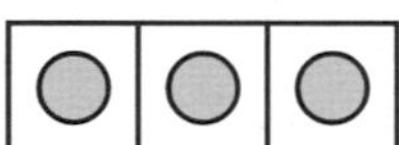

The PE department has
10 squash balls:
they need 4 boxes to put
the balls away.

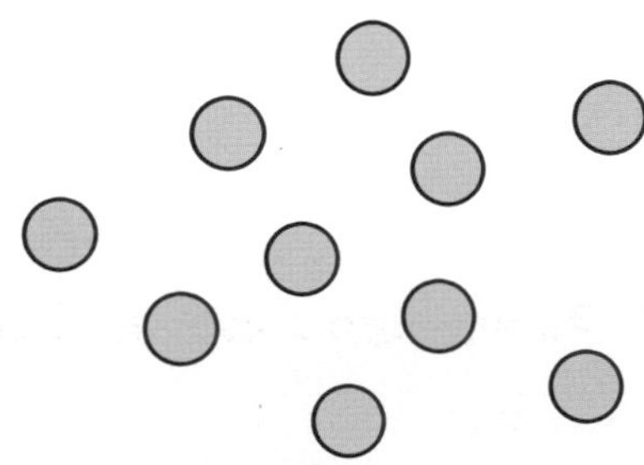

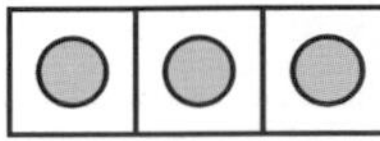 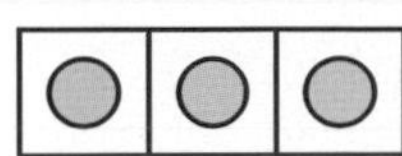

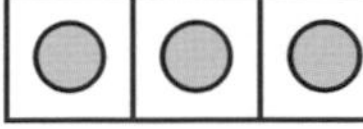 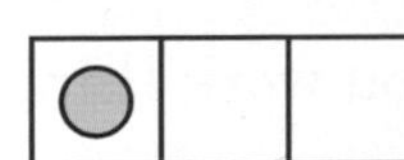

 2

2 How many boxes do they need for these squash balls?

a 5

b 8

c 11

d 17

e 20

f 14

This is a golf ball. They come in boxes of 10.

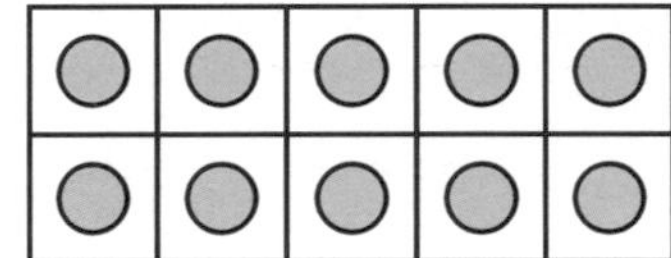

The PE department has 19 golf balls:

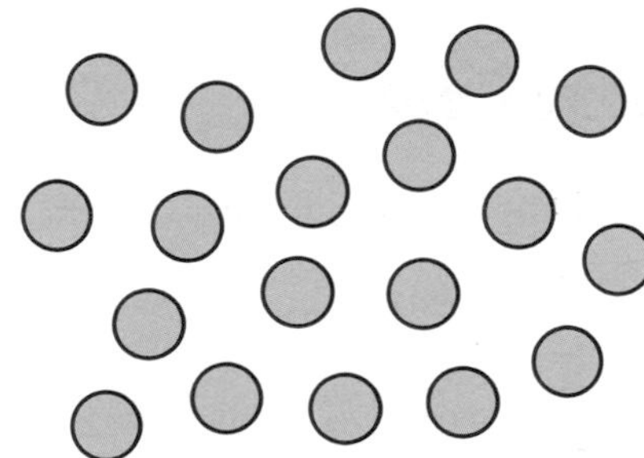

they need 2 boxes to put the balls away.

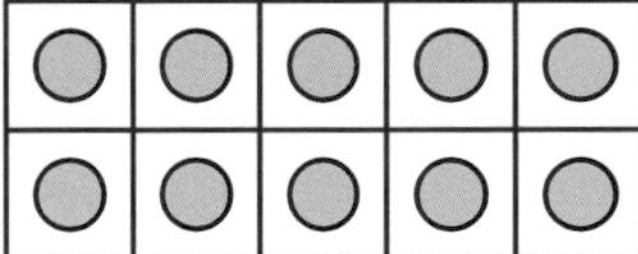

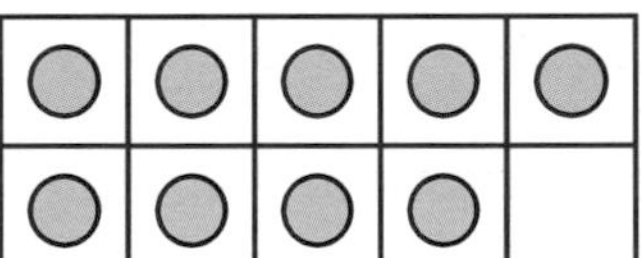

 3

3 How many boxes do they need for these golf balls?

a 5

b 24

c 31

d 17

e 50

4 **f** 48

2 Mental arithmetic

Exercise 4:5

 5

1 You need a times sheet.
Do not use a calculator.
Fill it in as quickly as you can.

2 Check your answers with a calculator.

3 You can use your times sheet to help you.
Do not use a calculator.
Write down the answers to these sums.

a $2 \times 2 = ?$	**d** $3 \times 2 = ?$	**g** $7 \times 2 = ?$
b $5 \times 2 = ?$	**e** $6 \times 2 = ?$	**h** $9 \times 2 = ?$
c $4 \times 2 = ?$	**f** $8 \times 2 = ?$	**i** $10 \times 2 = ?$

G 2

 5

4 Write these answers to the 2 times table in order, smallest first.

4 6 18 12 2 10 8 14 20 16

5 You can use your times sheet to help you.
Do not use a calculator.
Write down the answers to these sums.

a $2 \times 5 = ?$ **d** $9 \times 5 = ?$ **g** $3 \times 5 = ?$

b $4 \times 5 = ?$ **e** $6 \times 5 = ?$ **h** $5 \times 5 = ?$

G 2 **c** $7 \times 5 = ?$ **f** $8 \times 5 = ?$ **i** $10 \times 5 = ?$

H 6 **6** Write these answers to the 5 times table in order, smallest first.

H 7 10 15 30 45 5 25 20 35 50 40

7 You can use your times sheet to help you.
Do not use a calculator.
Write down the answers to these sums.

a $4 \times 10 = ?$ **d** $3 \times 10 = ?$ **g** $7 \times 10 = ?$

b $5 \times 10 = ?$ **e** $9 \times 10 = ?$ **h** $10 \times 10 = ?$

G 3 **c** $2 \times 10 = ?$ **f** $8 \times 10 = ?$ **i** $6 \times 10 = ?$

H 8 **8** Write the answers to the 10 times table in order, smallest first.

20 30 60 90 10 50 40 70 100 80

9 You can use your times sheet to help you.
Do not use a calculator.
Write down the answers to these sums.

a $2 \times 3 = ?$ **d** $3 \times 3 = ?$ **g** $8 \times 3 = ?$

b $7 \times 3 = ?$ **e** $9 \times 3 = ?$ **h** $10 \times 3 = ?$

G 3 **c** $4 \times 3 = ?$ **f** $6 \times 3 = ?$ **i** $5 \times 3 = ?$

9

10

4, 5, 6

10 Write the answers to the 3 times table in order, smallest first.

6 12 30 3 15 21 18 9 27 24

You can use times tables to divide.

Sarah, Jane and Melissa have two apples each.

They put them on the table:

$3 \times 2 = 6$

They share them out again.

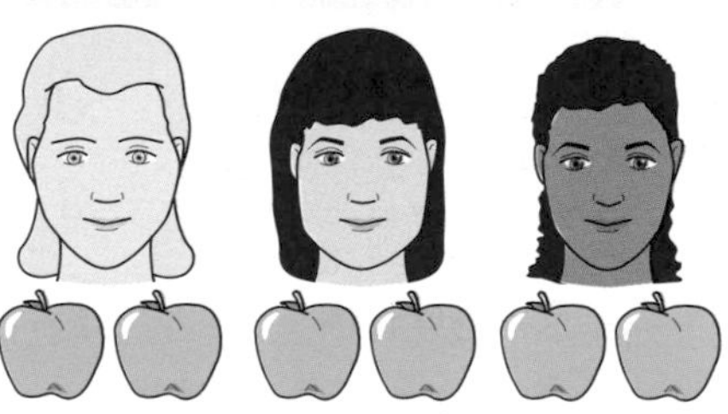

$6 \div 3 = 2$

Exercise 4:6

Use times tables to help with these sums.

1 **a** $2 \times 5 = 10$
$10 \div 2 = ?$

b $3 \times 4 = 12$
$12 \div 3 = ?$

c $5 \times 10 = 50$
$50 \div 5 = ?$

d $10 \times 4 = 40$
$40 \div 10 = ?$

e $2 \times 9 = 18$
$18 \div 2 = ?$

f $3 \times 3 = 9$
$9 \div 3 = ?$

3 Estimation

Estimating

John buys 1 chocolate bar for 8 p.
Here are the **estimated** costs of chocolate bars.

1 chocolate bar costs around 10 p
2 chocolate bars cost around 20 p

Exercise 4:7

1 Fergus buys 1 packet of mints for 19 p.
19 p rounded to the nearest 10 is 20 p.

Work out the estimated costs for 1 and 2 packets of mints.

2 Sonya buys a packet of chews for 7 p.

Work out the estimated costs for 1 and 2 packets of chews.

3 Andrew buys a packet of Bubble for 17 p.

Work out the estimated costs for 1 and 2 packets of Bubble.

4 Alexander buys a packet of crisps for 18 p.

Work out the estimated costs for 1 and 2 packets of crisps.

5 Jock buys a small packet of Fizzos for 6 p.

Work out the estimated costs for 1 and 2 packets of Fizzos.

Example Patrick buys a large packet of Fizzos for 9 p and a packet of Bubble for 17 p.
Find the estimated costs.

9 p

17 p

Estimated costs:	Fizzos	10 p
	Bubble	20 p
	Total	30 p

Exercise 4:8

Find the estimated costs.

1 Liam buys a chocolate bar for 8 p and a packet of crisps for 18 p.

8 p 18 p

2 Anna buys a packet of mints for 19 p and a packet of Bubble for 17 p.

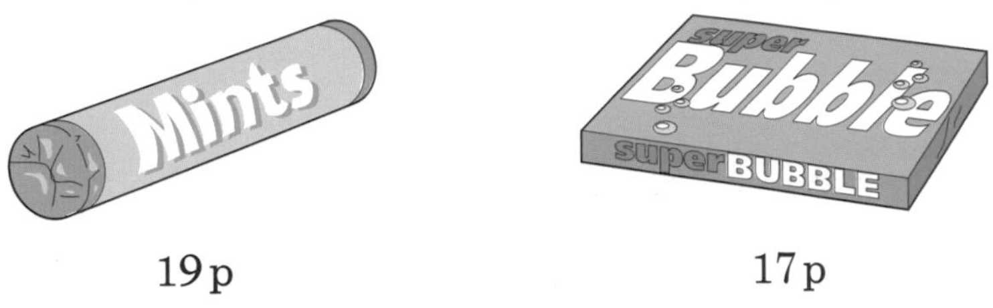

19 p 17 p

3 Melvin buys 2 packets of Fizzos which cost 6 p each and a packet of mints which costs 19 p.

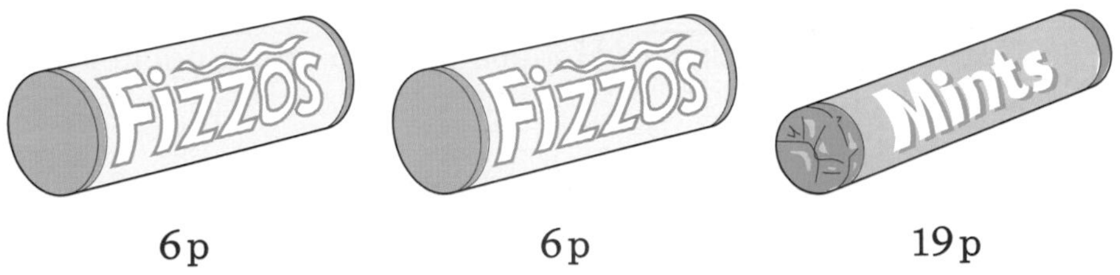

6 p 6 p 19 p

4 Clive buys 2 packets of chews for 7 p each and a packet of Fizzos for 9 p.

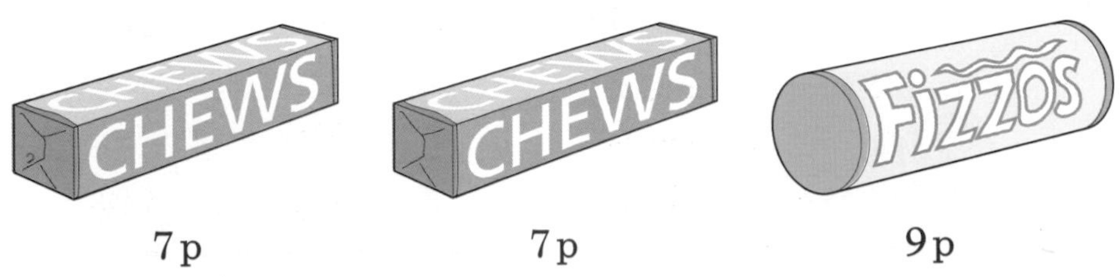

7 p 7 p 9 p

Using a calculator for accuracy

Estimating gives a rough answer.
Using a calculator carefully can be more accurate.

You need to **estimate** then **calculate**.

Example Fiona buys a packet of Munchers for 39 p and a packet of Loopies for 19 p.

Estimate:	Munchers	40 p
	Loopies	20 p
	Total	60 p

Calculate: 3 9 + 1 9 = 5 8

Total 58 p

Exercise 4:9

Estimate then calculate for each of these problems.

1 Carla buys a packet of Hoppos for 26 p and a packet of Zoomers for 39 p.

2 Petra buys a packet of Zimba for 27 p and a packet of Tappers for 19 p.

Mr Smith uses estimation to check if his pupil's calculations look correct.
This is Sarah's answer:

$48 + 11 + 21 = 100$

Mr Smith estimates
$50 + 10 + 20 = 80$
He knows her answer is wrong.

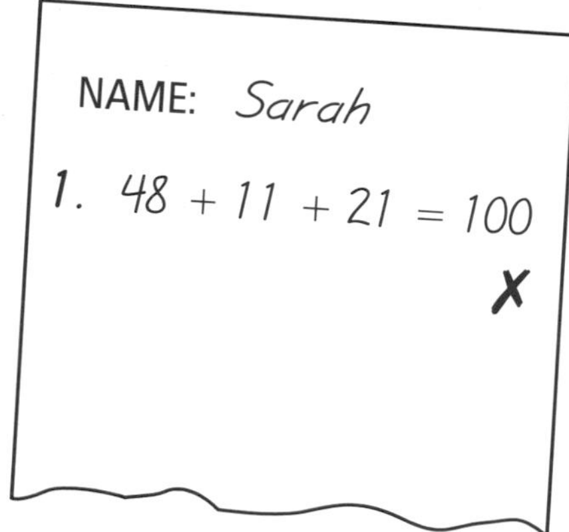

Exercise 4:10

Use estimation to find out if these calculations are right or wrong.

1 $52 + 9 + 41 = 102$

a Write down the estimated calculation.

b Write down right or wrong.

2 $11 + 24 + 16 = 81$

a Write down the estimated calculation.

b Write down right or wrong.

3 $39 + 22 + 14 + 29 = 124$

a Write down the estimated calculation.

b Write down right or wrong.

4 $9 + 72 + 34 + 59 = 174$

a Write down the estimated calculation.

b Write down right or wrong.

Division 1

Copy these into your book.
Work out the answers.

Exercise 1

1 $2 \div 2$

2 $12 \div 2$

3 $4 \div 2$

4 $14 \div 2$

5 $6 \div 2$

6 $16 \div 2$

7 $8 \div 2$

8 $18 \div 2$

9 $10 \div 2$

10 $20 \div 2$

Exercise 2

1 $2\overline{)8}$

2 $2\overline{)6}$

3 $2\overline{)2}$

4 $2\overline{)4}$

5 $2\overline{)10}$

6 $2\overline{)20}$

7 $2\overline{)12}$

8 $2\overline{)16}$

9 $2\overline{)18}$

10 $2\overline{)14}$

Exercise 3

1 $6 \div 3$

2 $15 \div 3$

3 $12 \div 3$

4 $24 \div 3$

5 $30 \div 3$

6 $27 \div 3$

7 $18 \div 3$

8 $9 \div 3$

9 $21 \div 3$

10 $3 \div 3$

Exercise 4

1 $3\overline{)6}$

2 $3\overline{)3}$

3 $3\overline{)24}$

4 $3\overline{)30}$

5 $3\overline{)9}$

6 $3\overline{)27}$

7 $3\overline{)21}$

8 $3\overline{)24}$

9 $3\overline{)18}$

10 $3\overline{)15}$

5 Shape and construction

CORE

QUESTIONS

HELP YOURSELF

Money 1

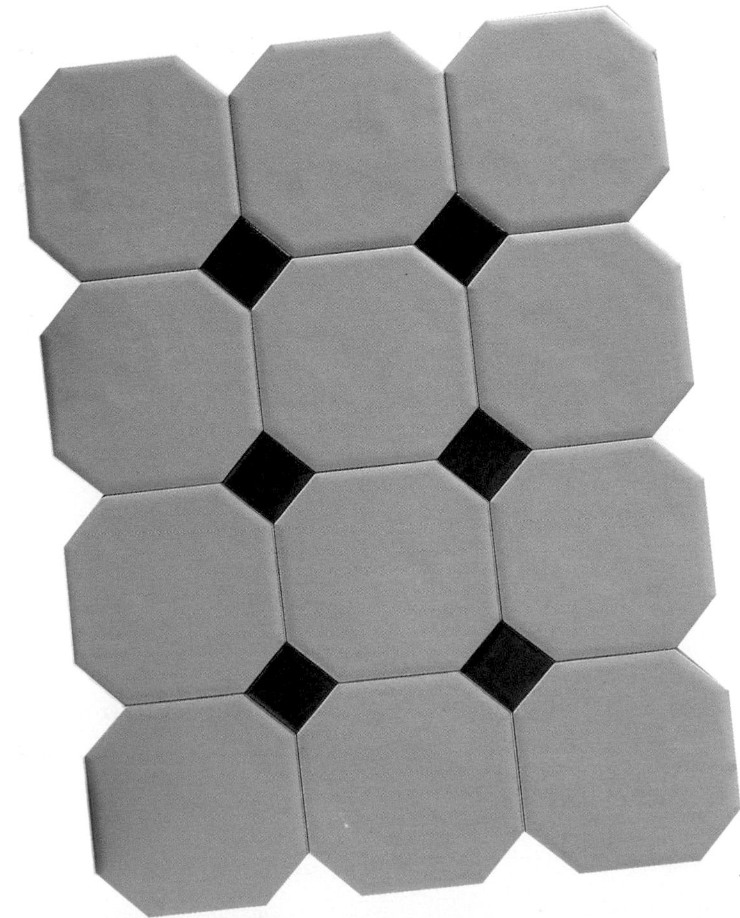

1 Names of polygons

Shapes can be seen everywhere. The design of this building uses shape for decoration.

Exercise 5:1

W1 **1** Sort the shapes below into 2 groups.
Group 1: Shapes with curved edges.
Group 2: Shapes with straight edges.

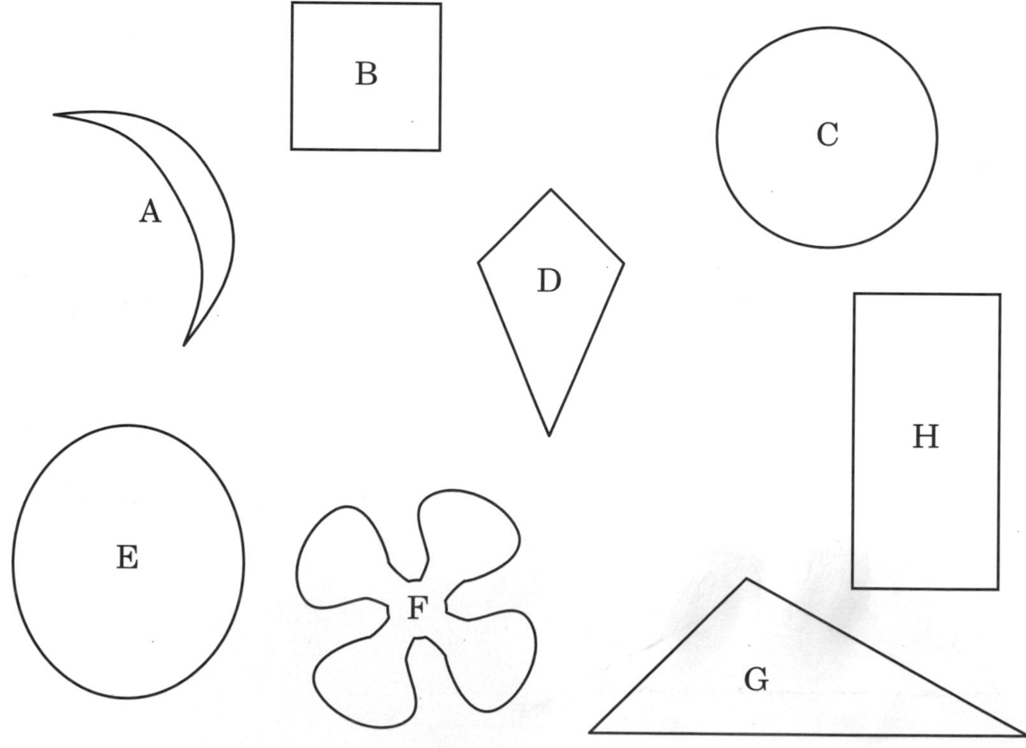

Polygon	A **polygon** is a shape with straight sides.

Exercise 5:2

W 2 **1** Sort the shapes below into 2 groups.
Group 1: Shapes with 3 sides.
Group 2: Shapes with more than 3 sides.

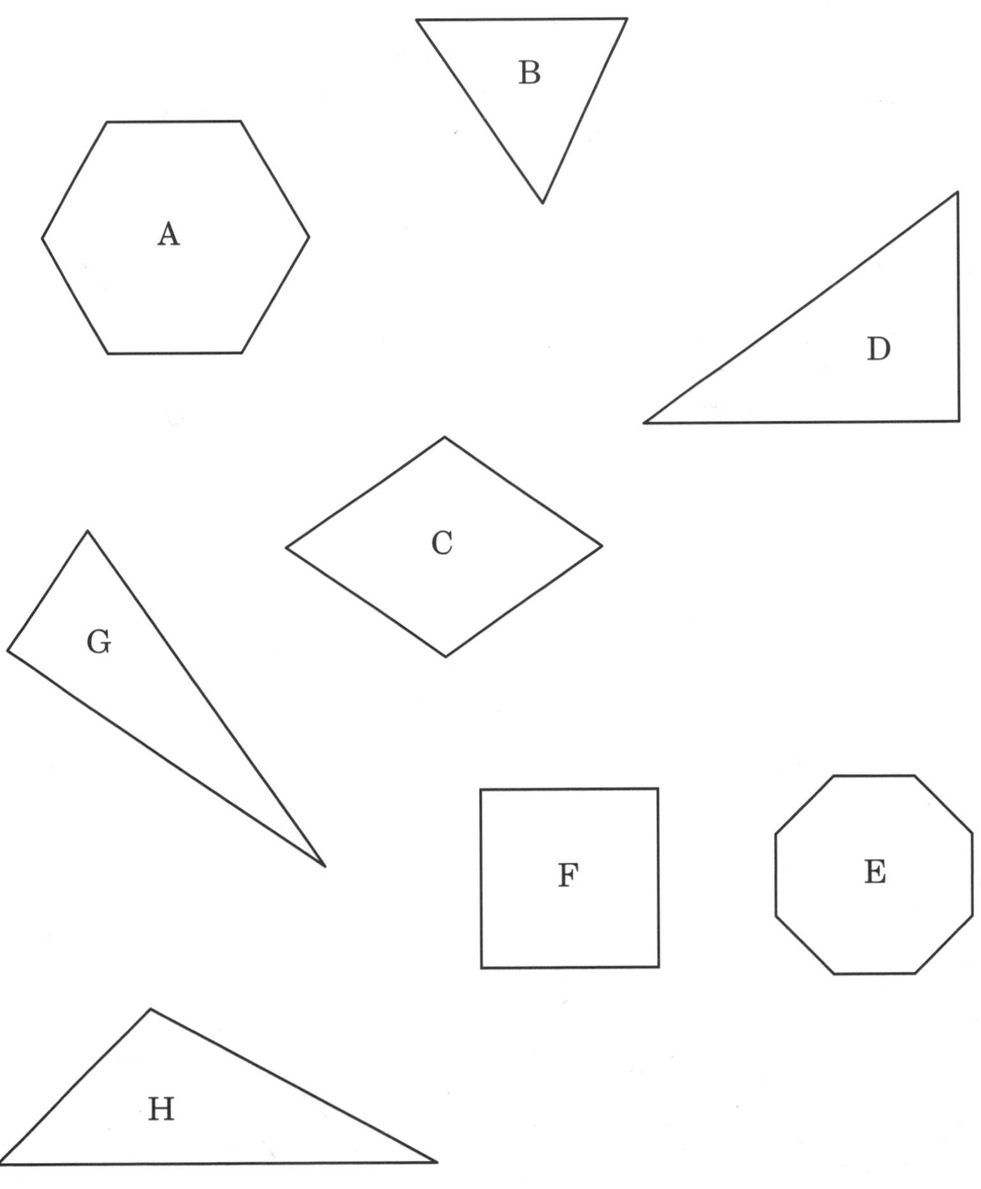

Triangle A shape with 3 sides is called a **triangle**.

Exercise 5:3

Use a ruler to measure the sides of each of these triangles.

1 Copy and complete.

$a = ?$ cm
$b = ?$ cm
$c = ?$ cm

The triangle has ? equal sides.

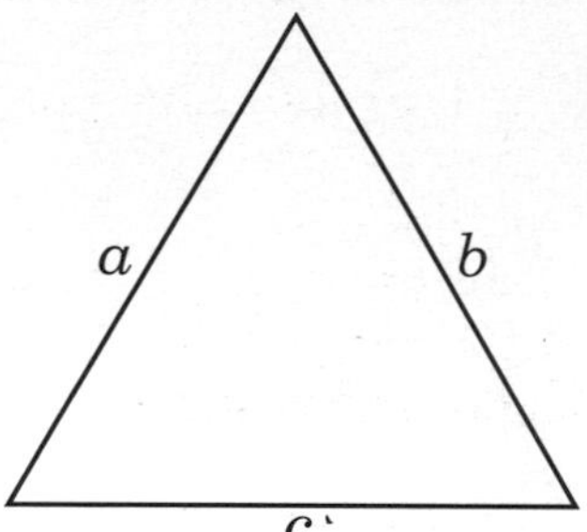

2 Copy and complete.

$d = ?$ cm
$e = ?$ cm
$f = ?$ cm

The triangle has ? equal sides.

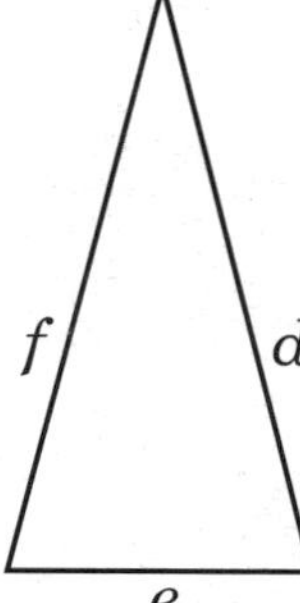

3 Copy and complete.

$g = ?$ cm
$h = ?$ cm
$j = ?$ cm

The triangle has ? equal sides.

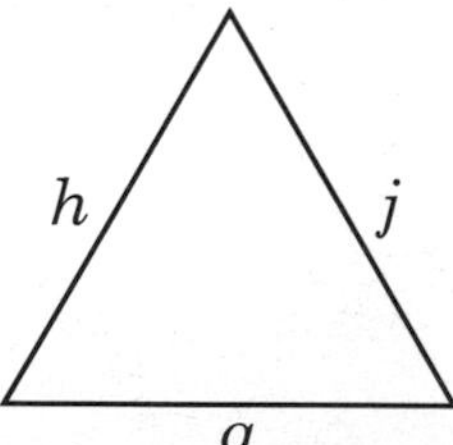

4 Copy and complete.

$k = ?$ cm
$l = ?$ cm
$m = ?$ cm

The triangle has ? equal sides.

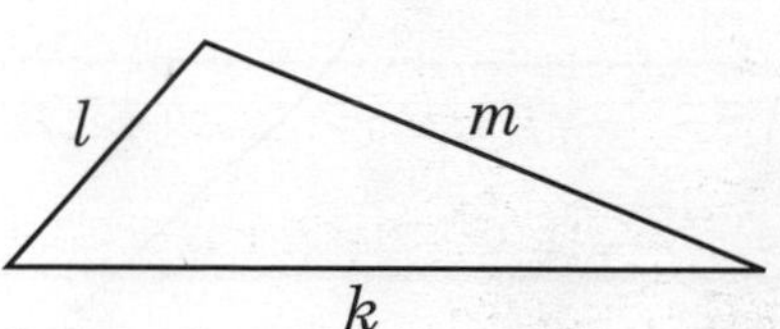

5 Copy and complete.

$n = ?$ cm
$p = ?$ cm
$q = ?$ cm

The triangle has ? equal sides.

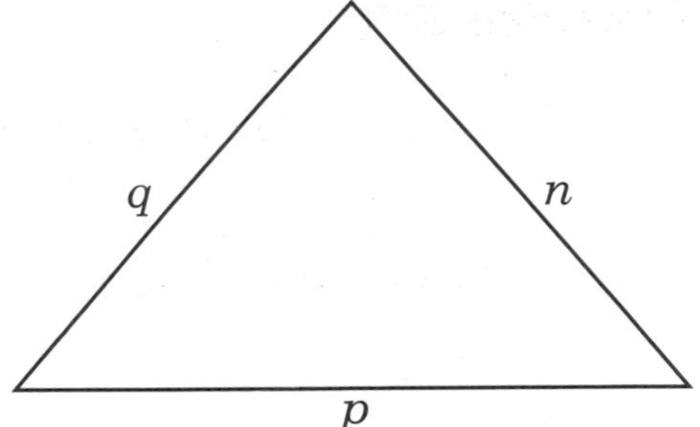

6 Copy and complete.

$r = ?$ cm
$s = ?$ cm
$t = ?$ cm

The triangle has ? equal sides.

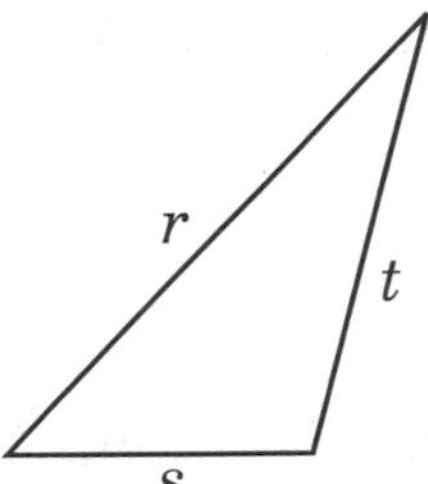

Equilateral triangle	An **equilateral triangle** has three equal sides.	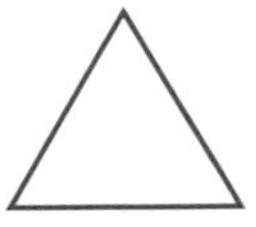
Isosceles triangle	An **isosceles triangle** has two equal sides.	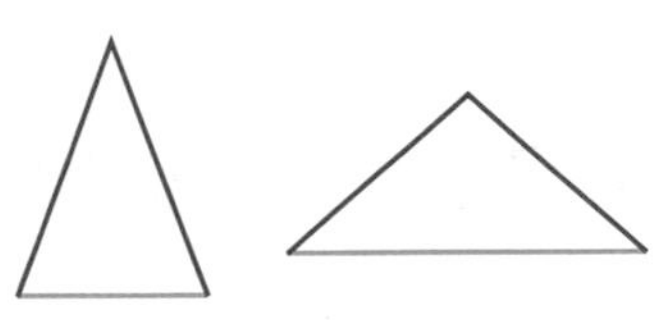
Scalene triangle	A **scalene triangle** has no equal sides.	

---

7 Look back at the triangles in Questions **1–6**.

a Which triangles are scalene?

b Which triangles are equilateral?

c Which triangles are isosceles?

 3 **8** You need copies of a clockface for this question.

a Join the points.

(1) 12 4 8 12
(2) 1 5 9 1
(3) 3 7 11 3

b What type of triangle are each of these?

c Join the points.

(1) 2 4 9 2
(2) 6 12 3 6
(3) 11 1 6 11
(4) 10 4 1 10

d What type of triangle are each of these?

Exercise 5:4

 4 **1** Sort these shapes into groups.
Group 1: Shapes with 4 sides.
Group 2: Shapes with 5 sides.
Group 3: Shapes with 6 sides.
Group 4: Shapes with 8 sides.

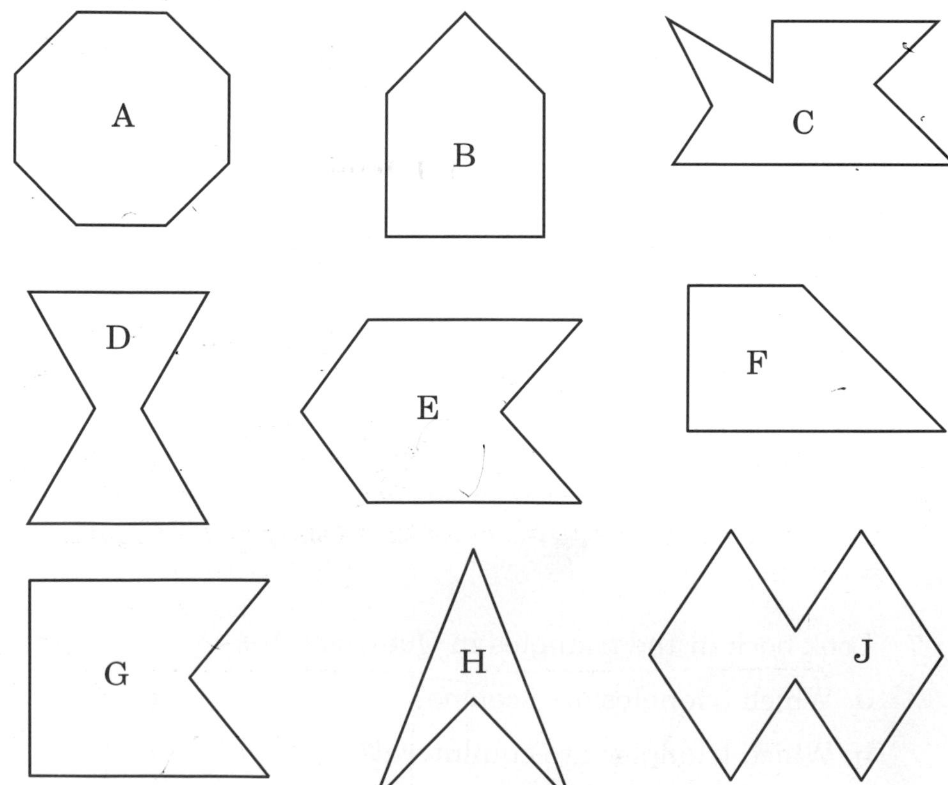

Quadrilateral A shape with 4 sides is called a **quadrilateral**.

Square **Rectangle** **Kite**

Trapezium **Parallelogram** **Rhombus**

Pentagon A shape with 5 sides is called a **pentagon**.

Hexagon A shape with 6 sides is called a **hexagon**.

Octagon A shape with 8 sides is called an **octagon**.

All of these shapes are polygons because they have straight sides.

Regular polygon In a **regular polygon** all the sides are the same length.
A square is a regular polygon.

Exercise 5:5

H 1

1 Use a ruler to find out which of these shapes are regular polygons.

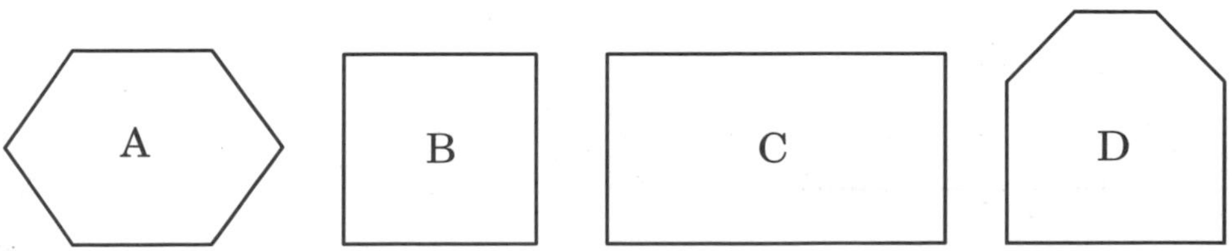

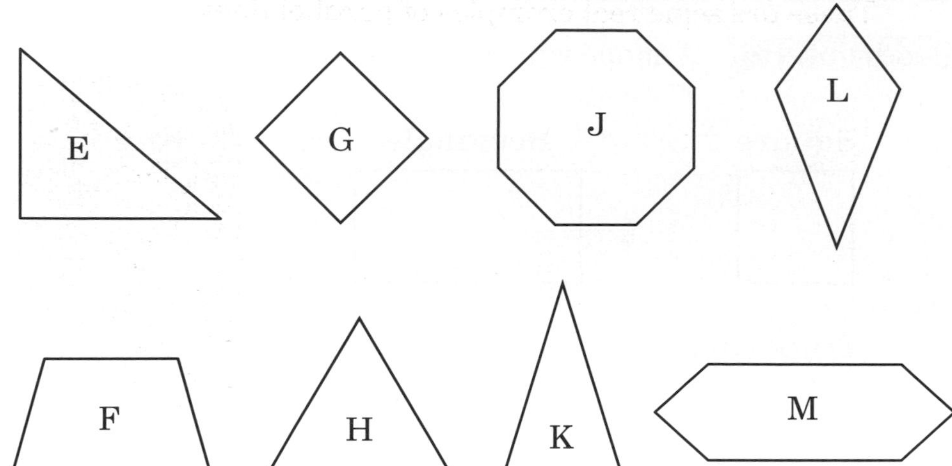

G 1, 2, 3, 4, 5, 6, 7

Exercise 5:6

 5

1 You will need copies of a clock face for this question.

a Join these points.

(1)	2	4	8	10	2
(2)	1	3	5	11	1
(3)	2	6	10	12	2
(4)	12	3	6	9	12
(5)	11	1	5	7	11
(6)	11	1	centre	9	11

b Label each shape with its special name.

c The colours on these shapes show sides of equal length.
Use colours to show sides of equal length on your shapes.

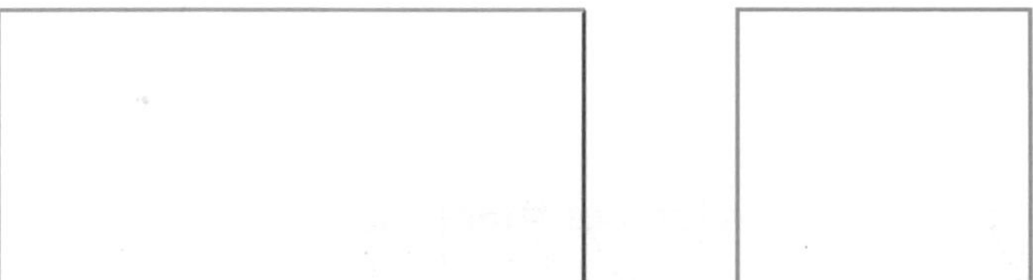

Parallel lines | **Parallel lines** never meet.
They stay the same distance apart.

These are some real examples of parallel lines:

Parallel lines do not have to be equal in length.

Exercise 5:7

1 Which of these pairs of lines are parallel?
You can use a ruler to measure them if you are not sure.

a

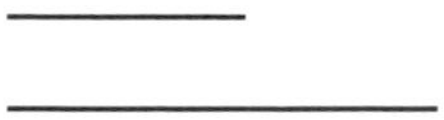

b 

c 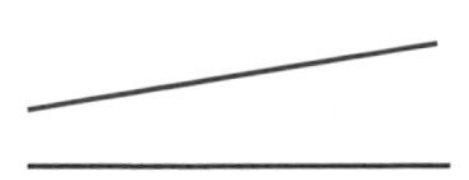

d 

e 

f 

6 **2** You need tracing paper for this question.
Trace the shapes below.
Use colour to show the pairs of parallel lines in each shape.
The first one has been done for you.

a

e

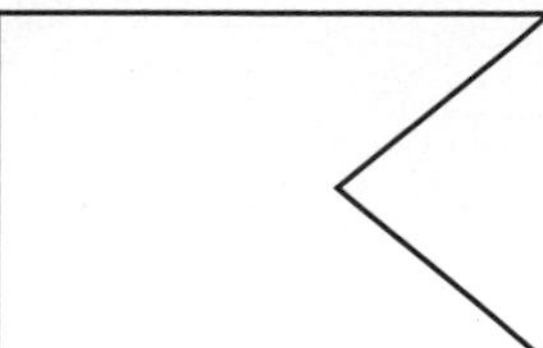

b

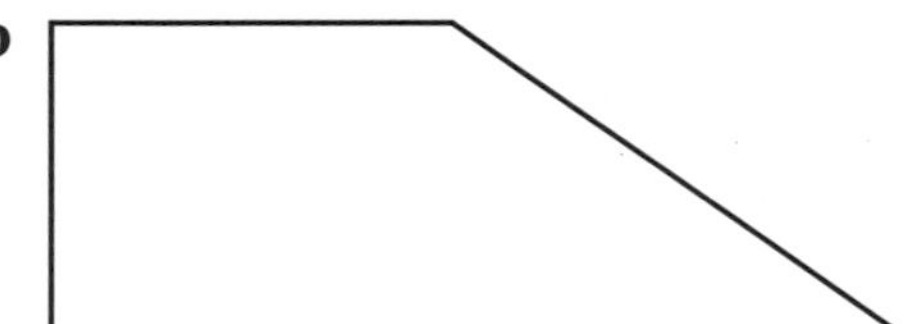

f

c

g

d

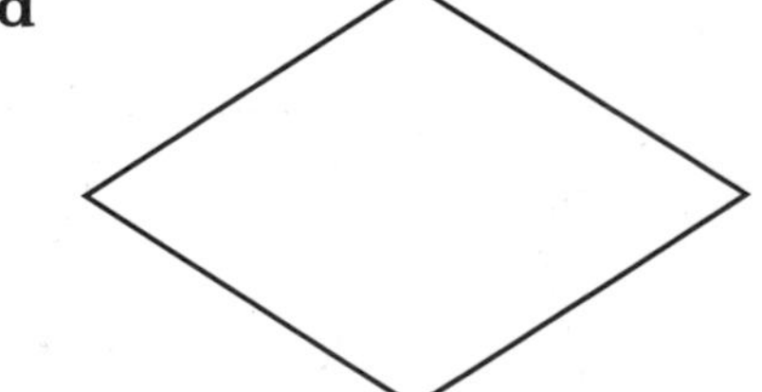

h

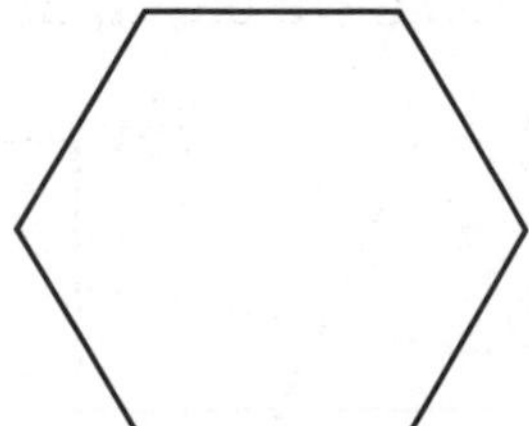

2 More about polygons

Vertex A point or corner of a shape is called a **vertex**.

Vertices For more than one point you say **vertices**.
A square has four vertices.

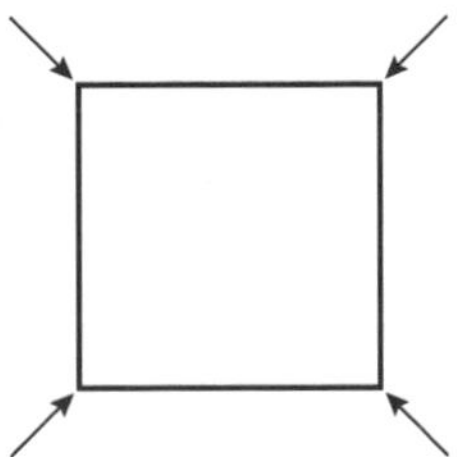

Exercise 5:8

W7 **1 a** Look carefully at these shapes.

(1) 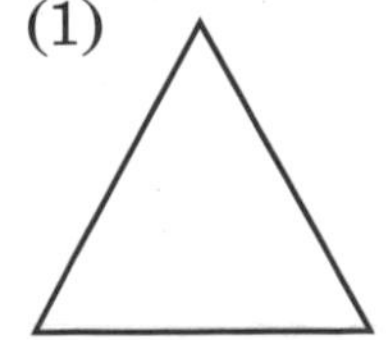 (2) 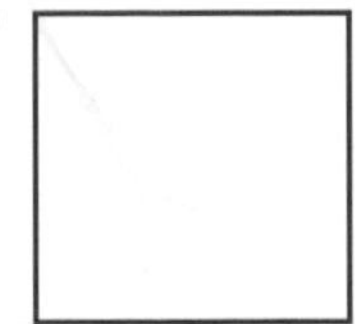 (3) 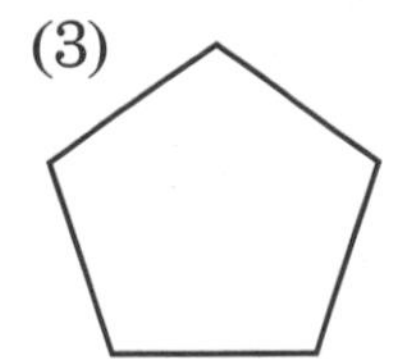 (4) 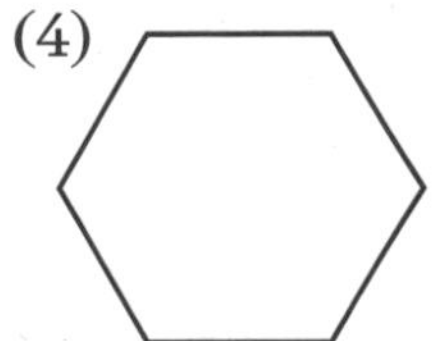

b Copy this table and fill it in.

Name of shape	Number of sides	Number of vertices	Number of diagonals
(1) triangle			
(2)		4	
(3)			5
(4)			

c What do you notice about the two columns with numbers in them?

W7 **2** How many vertices does an octagon have?

W7 **3** How many vertices does a rectangle have?

W7 **4** How many vertices does a 10-sided shape have?

Diagonal A line joining 2 vertices is called a **diagonal**.
One diagonal is drawn in this square.

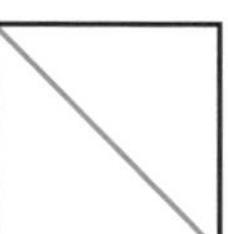

W8 **5** Trace these shapes.
Use a ruler to draw all the diagonals in each shape.
Make sure each vertex is joined to every other vertex.
The first one has been done for you.

a

b

c

d

e

f

g

h

5 Which of these trees are congruent?

6 Which of these boats are congruent?

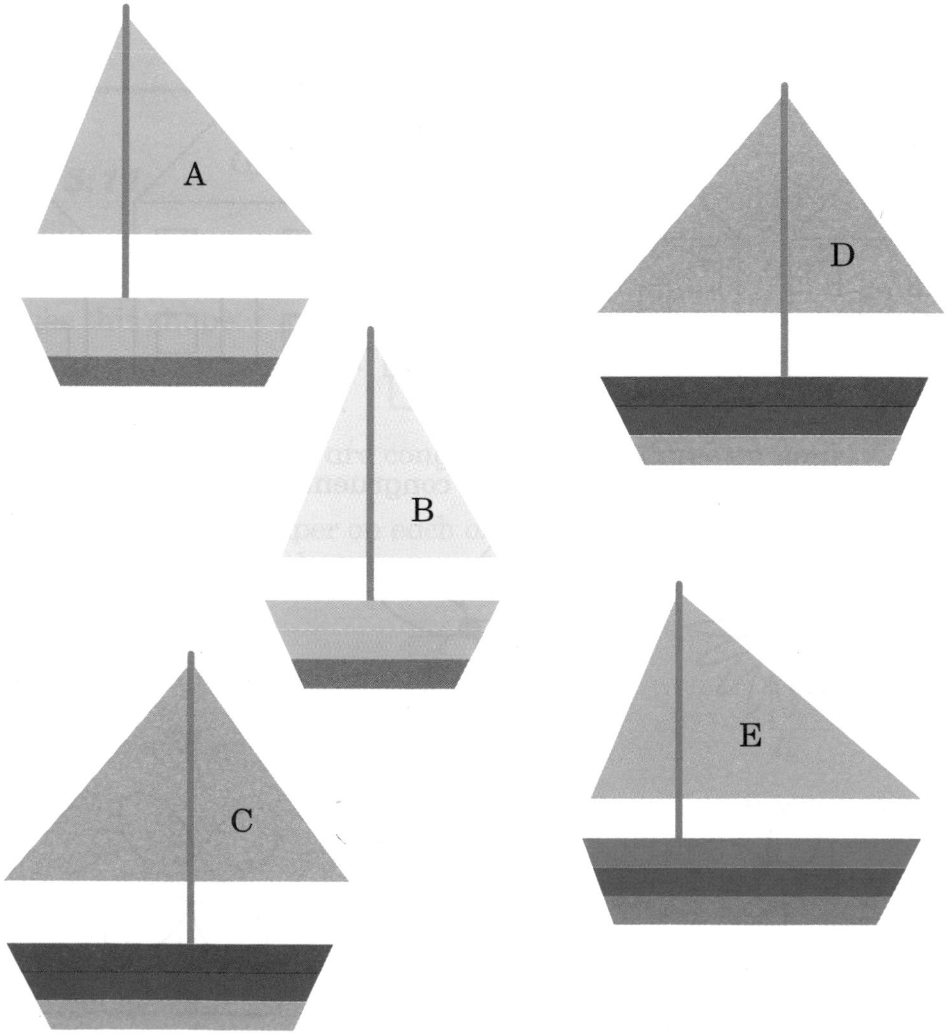

A

D

B

E

C

Tessellation A **tessellation** is a pattern made by repeating the same shape over and over again.

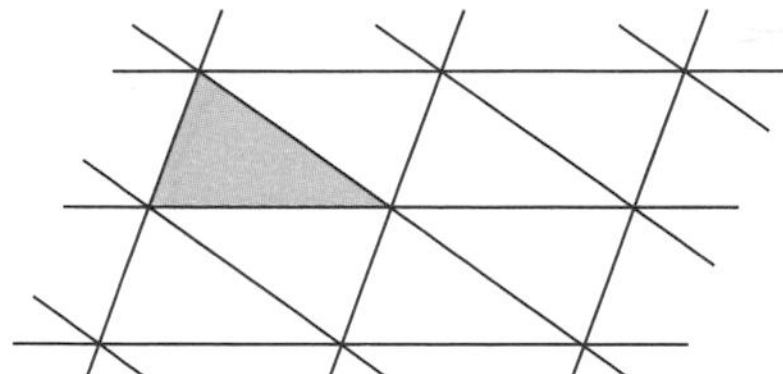

The triangle shape has been repeated over and over again.

Exercise 5:12

9

1 Mr Jones is building a patio. He uses square slabs.
Finish Mr Jones' patio with square slabs.

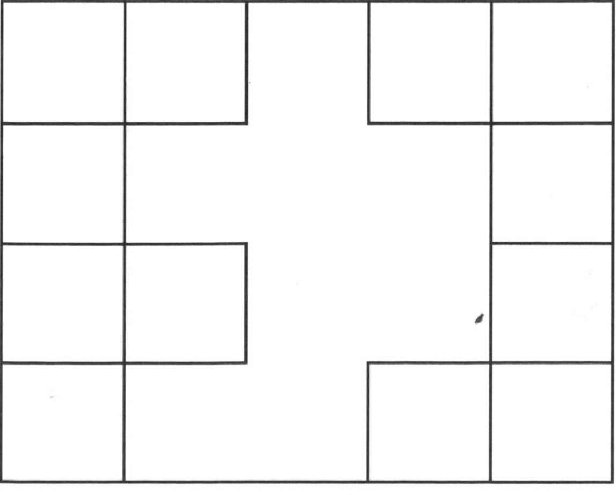

1 Sort these shapes into 3 groups:

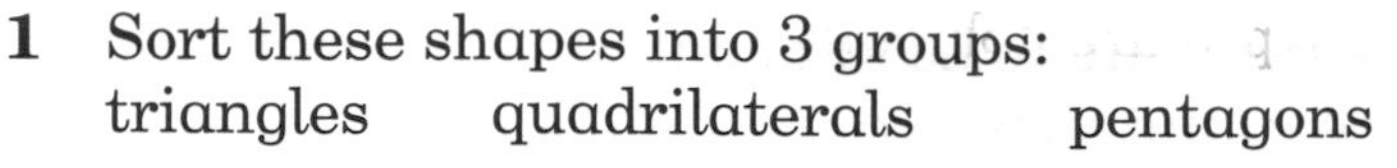

triangles quadrilaterals pentagons

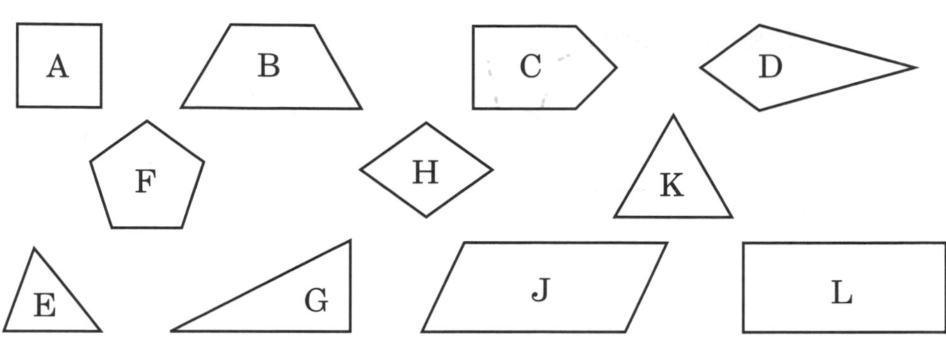

H 2

2 Warren has used a pin board to make 6 different triangles.
Write down the special name for each of the triangles:
equilateral, isosceles or scalene.

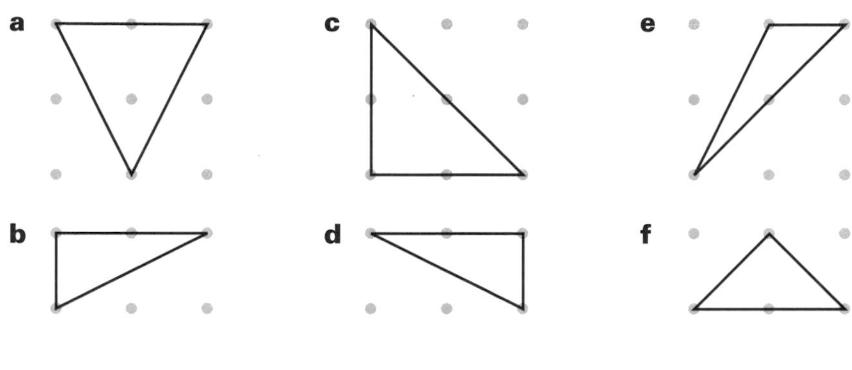

H 2

3 Charlene uses the pin board to make quadrilaterals.
Write down the special name for each quadrilateral:
square, rectangle, kite, parallelogram, rhombus or trapezium.

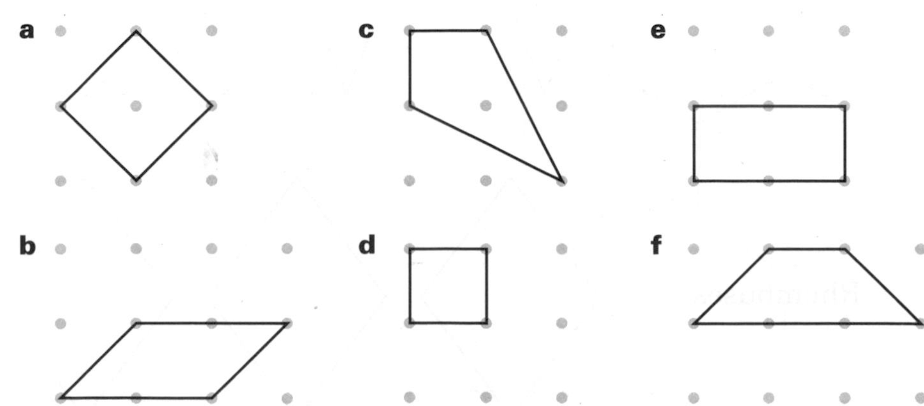

4 Which of these shapes are polygons?

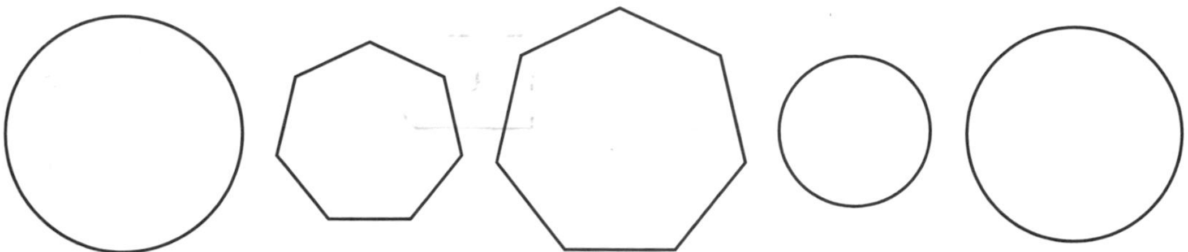

5 Use your compasses to draw a circle of radius 5 cm.

6 Which of these mugs are congruent?
You could use tracing paper to help you.

7 Use triangular grid paper to make a tessellation of trapeziums like the one below. You could use a stencil or tracing paper to help you.

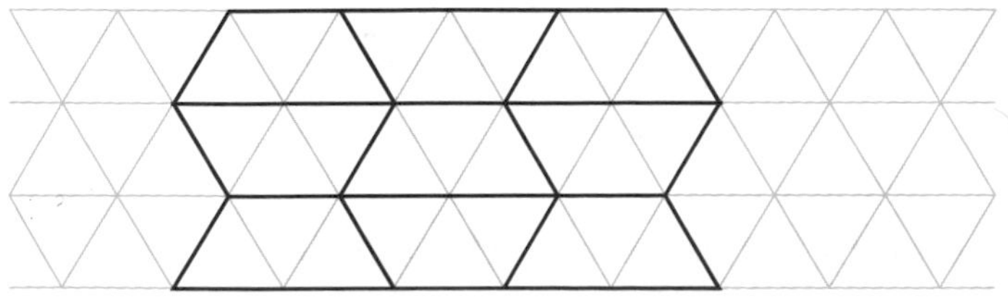

1 Place value in whole numbers

What number do these blocks show?

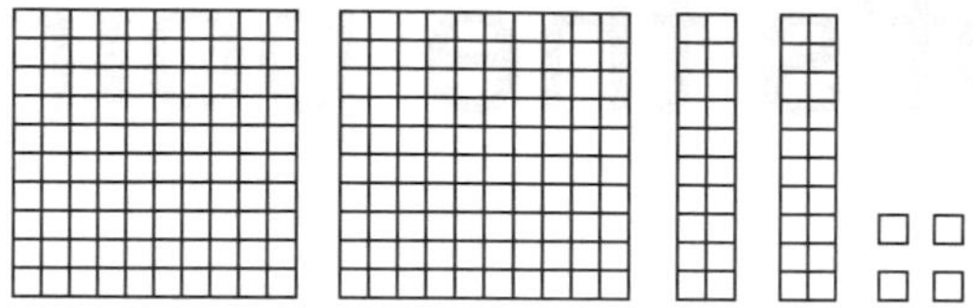

The blocks show **244** in figures.
This is **two hundred and forty-four** in words.

Exercise 6:1

H 1 **1** What do these blocks show:

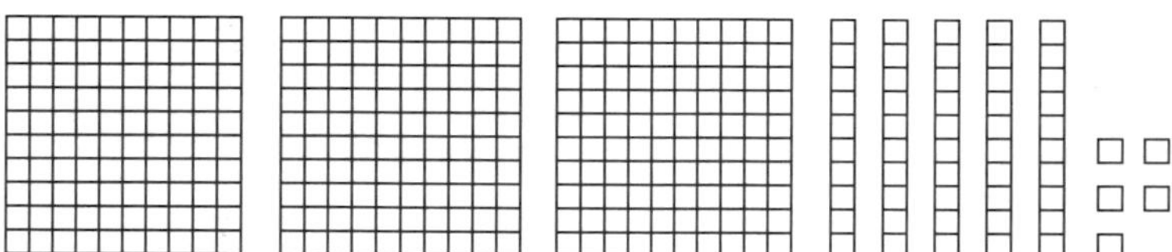

a in figures,

b in words?

H 1 **2** What do these blocks show:

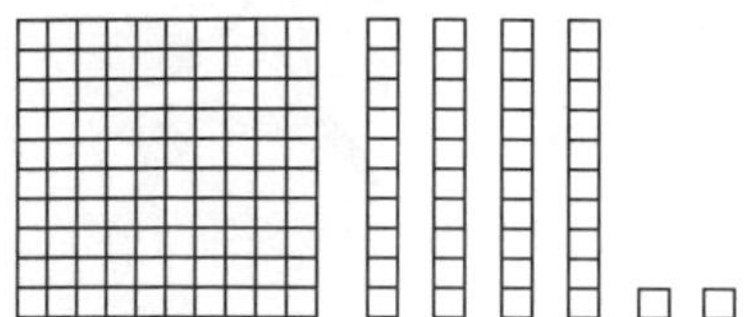

a in figures,

b in words?

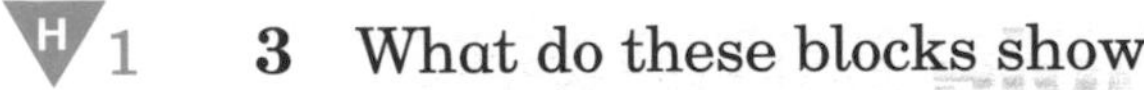

1 **3** What do these blocks show:

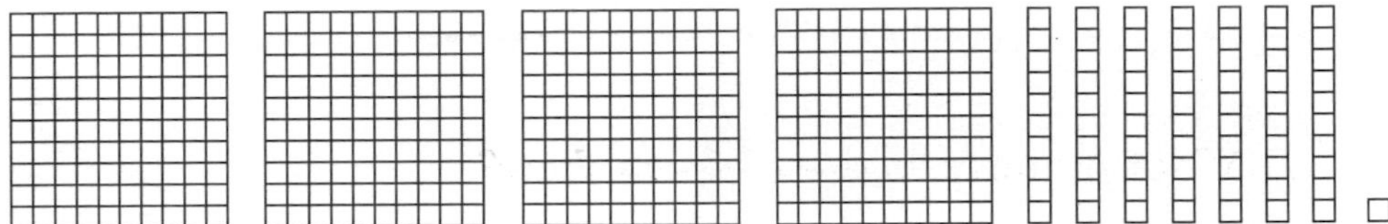

a in figures,

b in words?

You can write the number five hundred and seventy-four as 574 using the digits, 5, 7 and 4.

---

1 **4** Write these numbers in figures.

a Sixty-three

b Seven hundred and fifty-six

c Twenty-eight

d Four hundred and eleven

e Ninety-nine

f Fifteen

1 **5** Write these numbers in figures.

a Eighty-four

b Nineteen

c Seven hundred and twelve

d Thirty-nine

You can write **745** as **seven hundred and forty-five**.

H 1 **6** Write these numbers in words.

a 322 **d** 46

b 512 **e** 131

c 85 **f** 19

When you write a cheque you need to write the amount in words and figures.

KEY MATHS

Date 01/01/98

Pay Key Maths

The sum of Two hundred and thirty-two pounds

£ 232.00

Signed A. N. Other

W 1 **7** Fill in the cheques to show these amounts in words and figures.

a £567 **d** £285

b £42 **e** £670

c £313 **f** £898

W 1 **8** Fill in the cheques to show these amounts in words and figures.

a Two hundred and twenty pounds

b Seventy-eight pounds

c One hundred and six pounds

d Four hundred and forty-two pounds

e Six hundred and fifteen pounds

f Twenty-seven pounds

W 1 **9** Fill in the cheques to show these amounts in words and figures.

a Three hundred and thirty pounds

b Eighty-nine pounds

c Five hundred and seventy-three pounds

d Seven hundred and sixteen pounds

e Twenty-two pounds

f Forty-six pounds

Numbers are written in columns.
The column that the digit is in tells you its value.

Example The number 574 is on a place value grid.
What is the value of the 4?

Hundreds 100s	Tens 10s	Units 1s
5	7	**4**

The **4** is **4 units.**

Exercise 6:2

 2 You can use a place value grid and a set of 0–9 number cards for this exercise.

1 What is the value of the **8** in 16**8**?

2 What is the value of the **6** in 21**6**?

3 What is the value of the **7** in 39**7**?

4 What is the value of the **1** in 52**1**?

5 What is the value of the **3** in 84**3**?

6 What is the value of the **4** in 98**4**?

Example The number 574 is on a place value grid. What is the value of the 7?

Hundreds 100s	Tens 10s	Units 1s
5	7	4

The **7** is **7 tens** or **70**.

---

7 What is the value of the **6** in 4**6**5?

8 What is the value of the **2** in 1**2**3?

9 What is the value of the **9** in 5**9**8?

10 What is the value of the **1** in 2**1**8?

11 What is the value of the **5** in 6**5**3?

12 What is the value of the **3** in 5**3**6?

Example The number 574 is on a place value grid. What is the value of the 5?

Hundreds 100s	Tens 10s	Units 1s
5	7	4

The 5 is **5 hundreds** or 500.

13 What is the value of the **3** in 386?

14 What is the value of the **8** in 875?

15 What is the value of the **3** in 379?

16 What is the value of the **2** in 201?

17 What is the value of the **7** in 718?

18 What is the value of the **6** in 645?

Exercise 6:3

1 What is the value of the **5** in

a 5 **b** 50 **c** 500?

2 What is the value of the **8** in

a 8 **b** 80 **c** 800?

3 What is the value of the **2** in

a 2 **b** 20 **c** 200?

4 What is the value of the red digit in each of these?

a 693 **c** 84 **e** 47

G 1 **b** 218 **d** 581 **f** 103

W 3 **2** Colour in these decimals on the worksheet.

a 0.2 **c** 0.7 **e** 0.9

b 0.5 **d** 0.6 **f** 0.4

Example There is 1 whole circle that is coloured and 0.4 of another circle.
How much is coloured altogether?

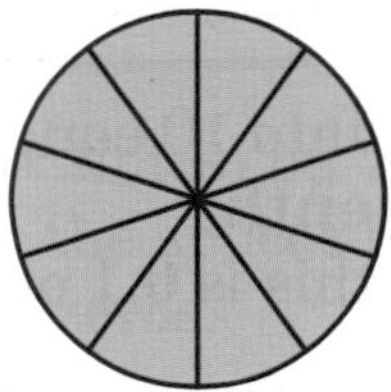 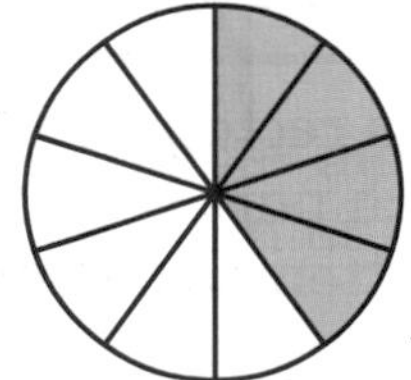

$1 + 0.4 = 1.4$

The total amount that is coloured is 1.4.

---

3 In each question, write down the total amount that is coloured.

a

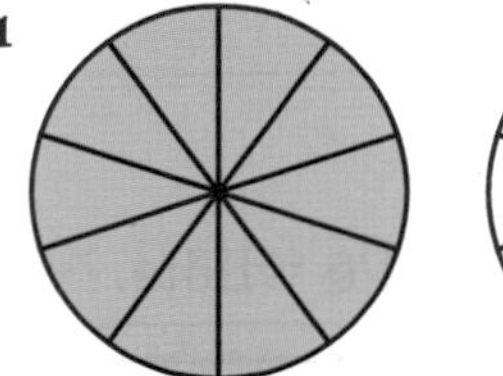 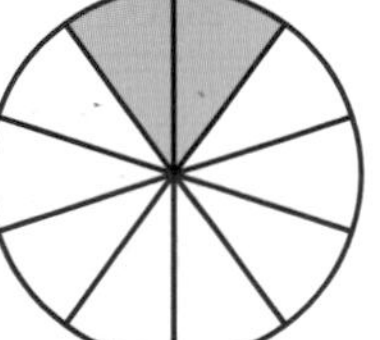

d

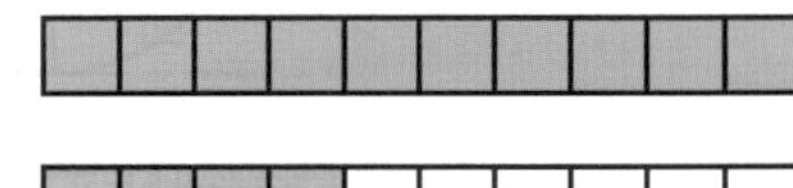

b

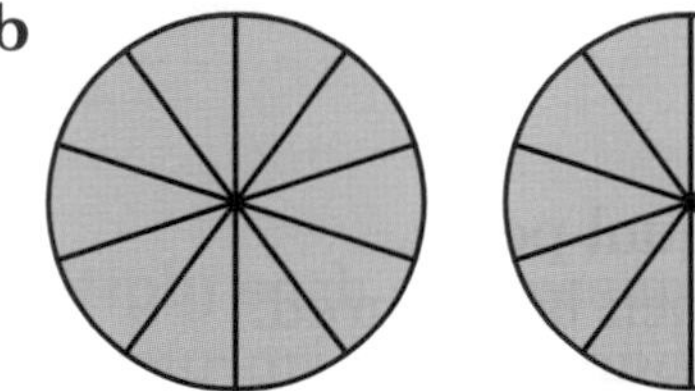

e

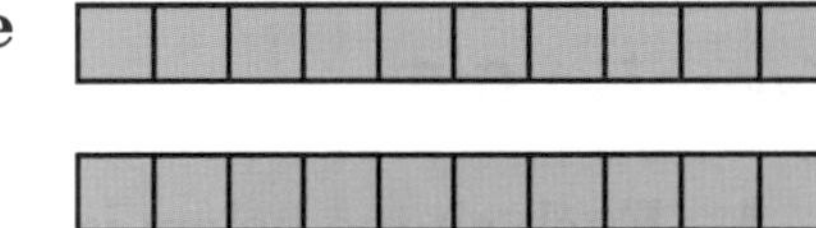

c

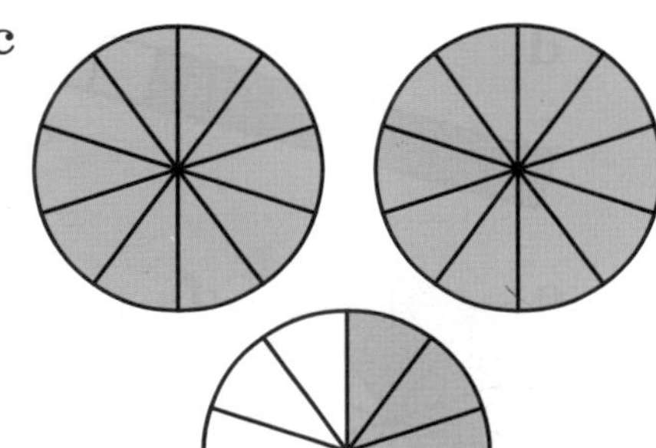

f

Example Here is 0.5 of a rectangle and 0.2 of another rectangle. What is the total amount that is coloured?

If you add them together this is the same as 0.7 of a rectangle.

0.5 + 0.2 = 0.7 

W 4 **4** In each question, write down the total amount that is coloured.

a

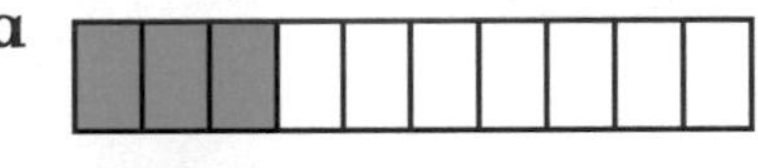

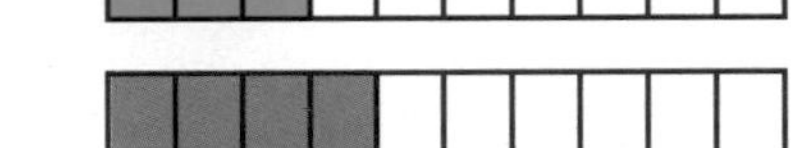

b

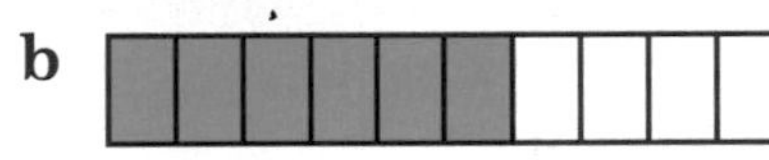

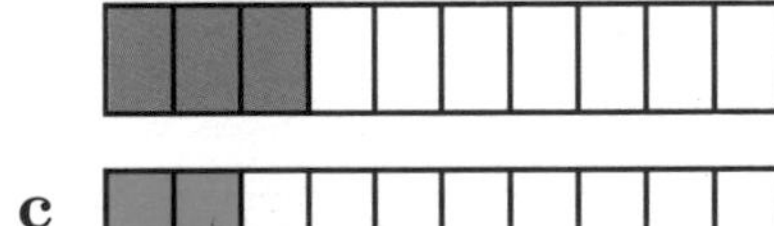

c

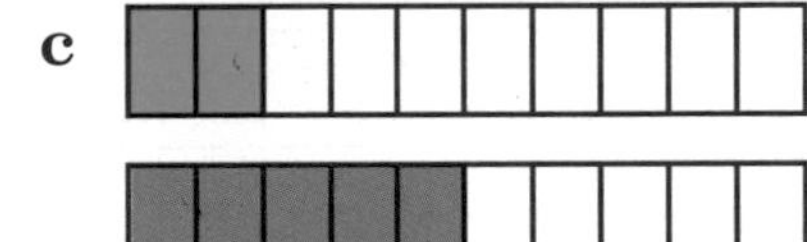

d

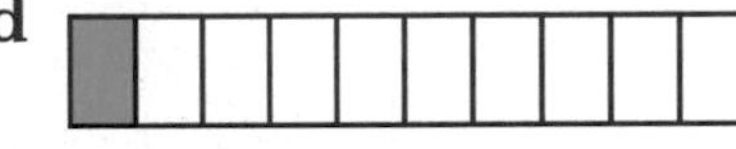

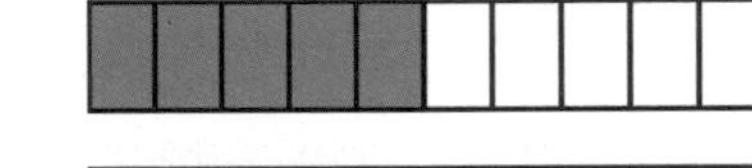

e

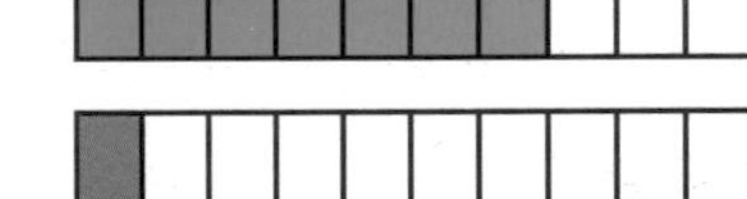

f

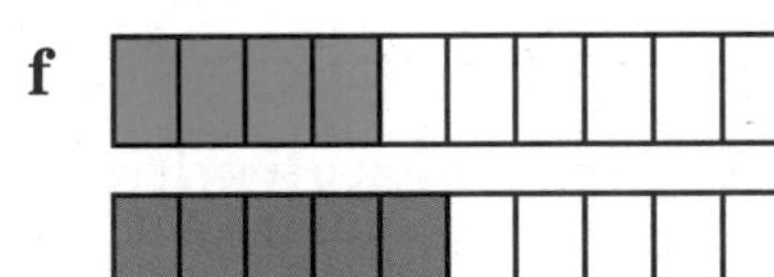

Example Here is 0.8 of a whole circle and 0.7 of another circle. What is the total amount that is coloured?

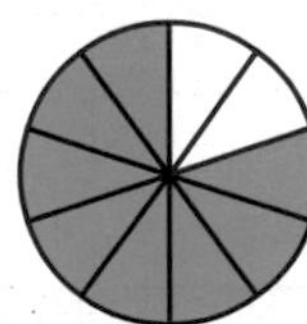 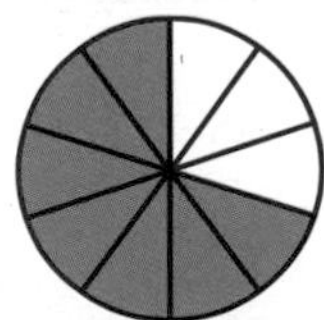

If you add them together this is the same as 1.5 of a circle.

0.8 + 0.7 = 1.5 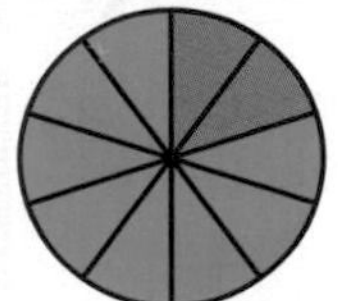 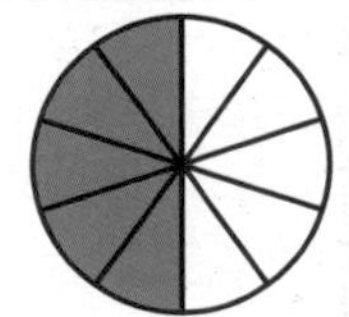

W 5 **5** In each question, write down the total amount that is coloured.

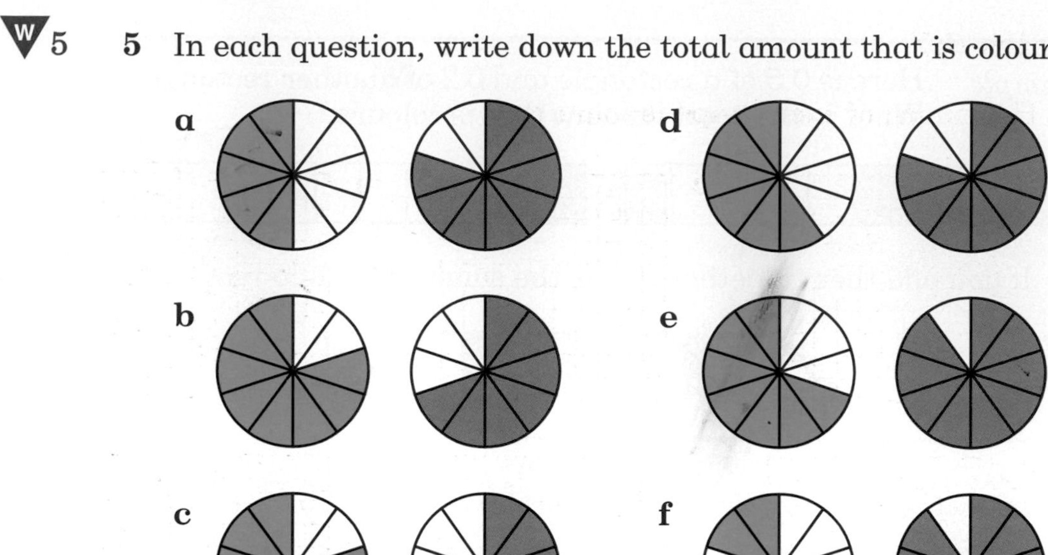

G 6

Hundredths

Hundredths **Hundredths** come from splitting 1 into 100 equal parts.
This square is split into 100 equal parts.

Each part is called a **hundredth.**

Written as a decimal this is **0.01** of a whole.

Example This square is split into 100 equal parts.

How much of the square is coloured?
35 of the parts are coloured.
Written as a decimal this is 0.35 of a whole.

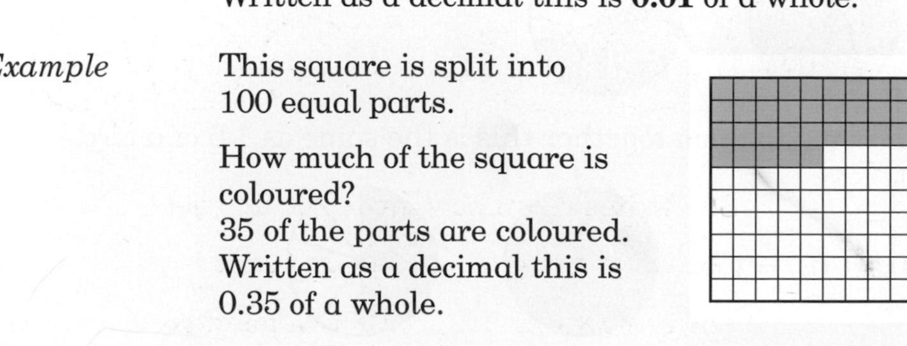

Exercise 6:7

1 How much of each shape is coloured?

a

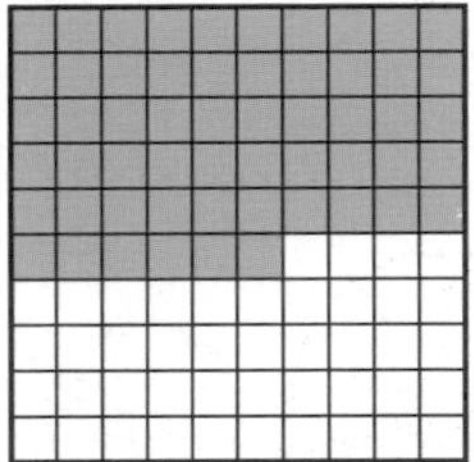

c

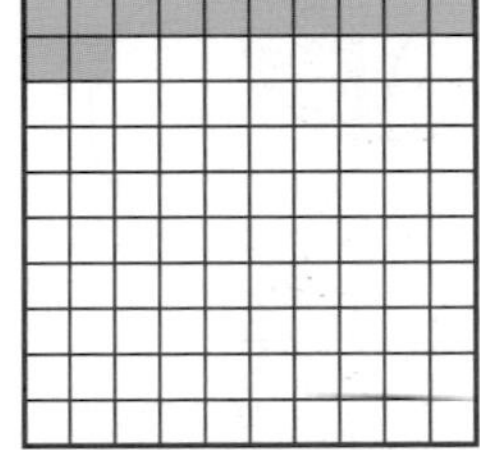

e

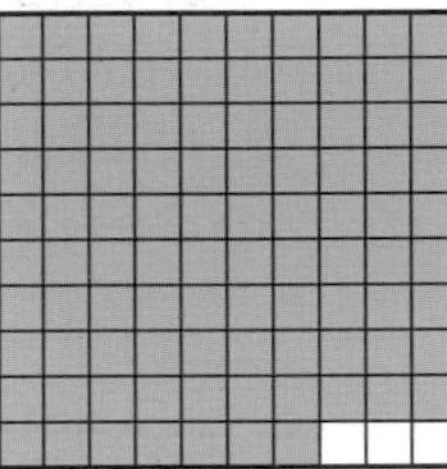

b

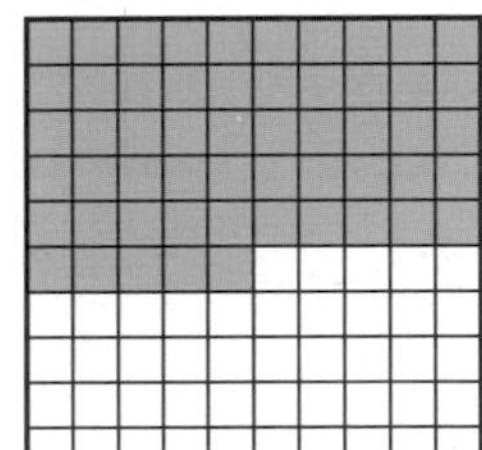

d

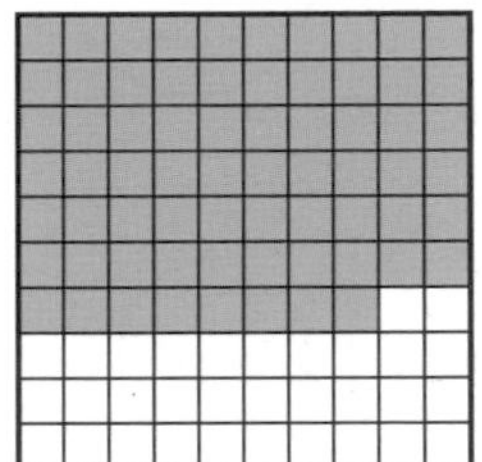

f

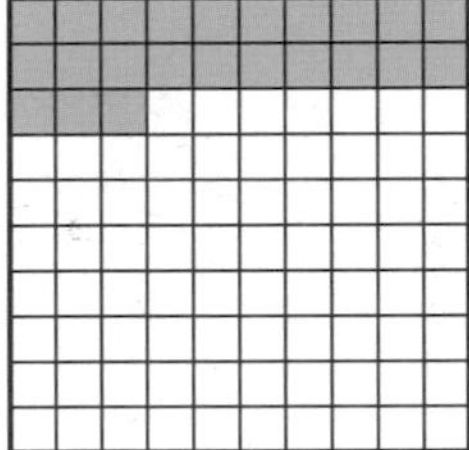

W6 **2** Colour in these decimals on your worksheet.

a 0.65 **d** 0.91

b 0.18 **e** 0.27

c 0.32 **f** 0.74

Decimals and money

Sarah has some money.

50 p

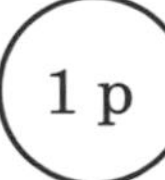

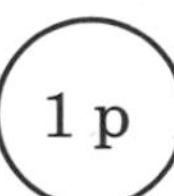

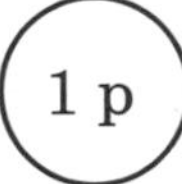

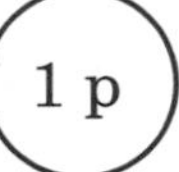

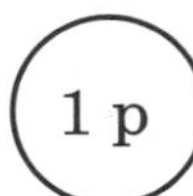

She writes down how much she has got.

She could write **56 p** in pence

She could write **£0.56** in pounds

Exercise 6:8

1 Write these amounts in pence.

a 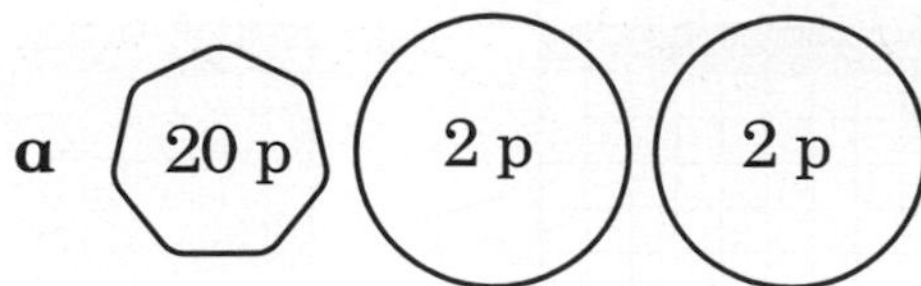

b 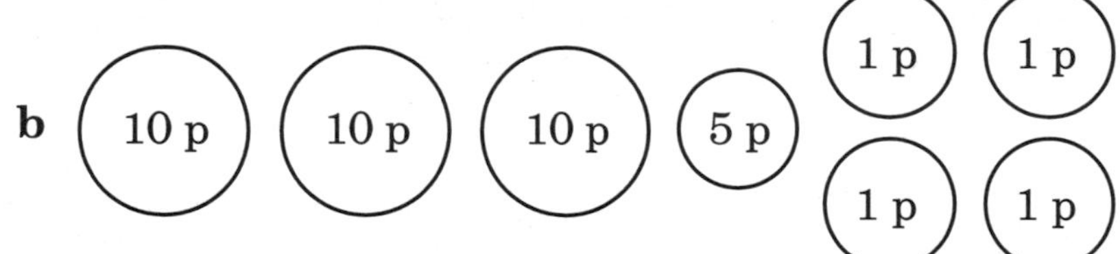

c 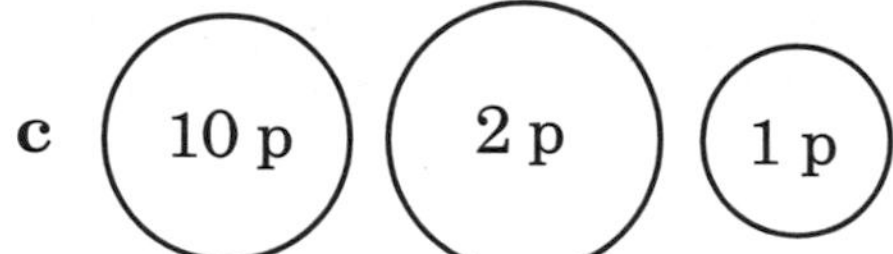

d 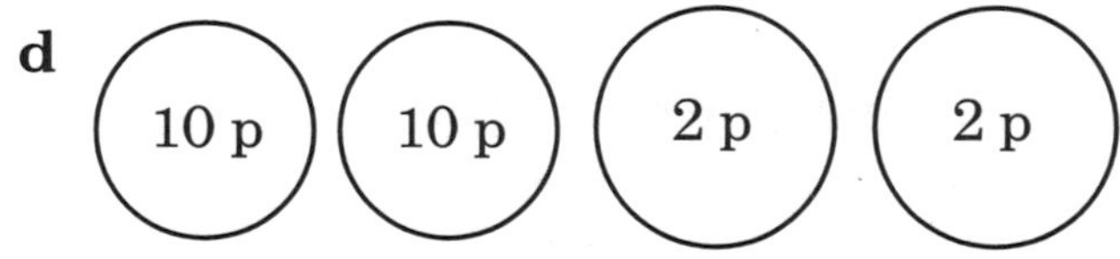

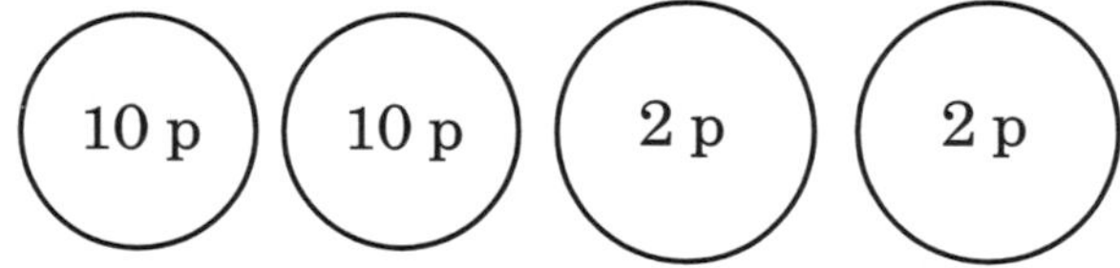

e 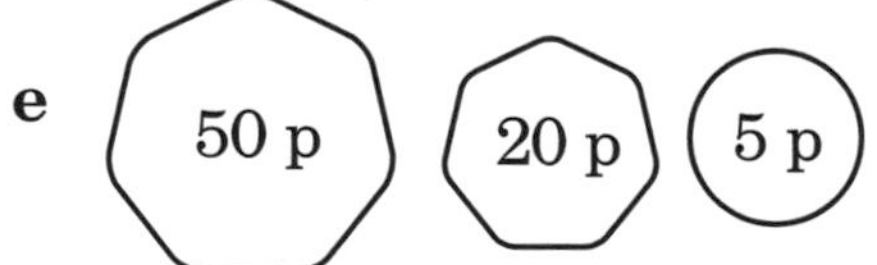

f 50 p 20 p 20 p 1 p

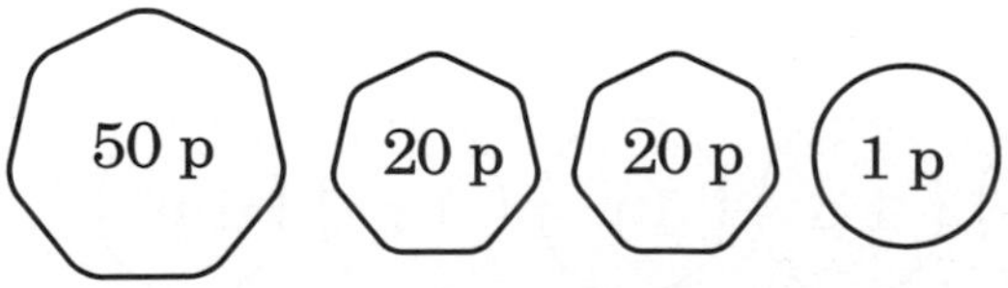

2 Write these amounts in pence.

a £0.42 **d** £0.64

b £0.93 **e** £0.77

c £0.36 **f** £0.19

3 Write these amounts in pounds.

a 10 p 10 p 5 p 1 p 1 p 1 p

b 20 p 10 p 2 p 1 p

c 10 p 10 p 10 p 5 p 1 p 1 p 1 p

d 10 p 5 p

e 50 p 5 p 2 p

f 20 p 20 p 20 p 20 p 1 p

4 Write these amounts in pounds.

a 65 p **d** 37 p

b 18 p **e** 92 p

c 21 p **f** 86 p

5 Write these amounts in pounds.

a 24 p **d** 88 p

b 72 p **e** 91 p

G 7 **c** 43 p **f** 67 p

Example Add up these 2 amounts of money.

26 p + 14 p

$$\begin{array}{r} 26\,\text{p} \\ +\ \underline{14\,\text{p}} \\ \underline{40\,\text{p}} \\ {\scriptstyle 1} \end{array}$$

You get 40 p

You know that 26 p in pence can be written as £0.26 in pounds.
You know that 14 p in pence can be written as £0.14 in pounds.

The calculator sum to add these 2 amounts of money is

0 **.** **2** **6** **+** **0** **.** **1** **4** **=**

You will see 0.4 on your calculator.

The calculator does not realise that you are using money and so you have to remember to put a zero on the end for the answer to make sense.

The answer is 40 p in pence or £0.40 in pounds.

Exercise 6:10

Use your calculator to help you find the answers to these questions.
Check your answers by doing the sums in your book.

1 35 p + 15 p

a Write down the calculator sum.

b Write down the answer on your calculator.

c Write your answer in pounds.

d Write your answer in pence.

2 21 p + 69 p

a Write down the calculator sum.

b Write down the answer on your calculator.

c Write your answer in pounds.

d Write your answer in pence.

3 18 p + 12 p

a Write down the calculator sum.

b Write down the answer on your calculator.

c Write your answer in pounds.

d Write your answer in pence.

4 53 p + 27 p

a Write down the calculator sum.

b Write down the answer on your calculator.

c Write your answer in pounds.

d Write your answer in pence.

5 42 p + 28 p

a Write down the calculator sum.

b Write down the answer on your calculator.

c Write your answer in pounds.

d Write your answer in pence.

6 23 p + 67 p

a Write down the calculator sum.

b Write down the answer on your calculator.

c Write your answer in pounds.

d Write your answer in pence.

7 42 p + 38 p

a Write down the calculator sum.

b Write down the answer on your calculator.

c Write your answer in pounds.

d Write your answer in pence.

Time 1

Write the times shown on these clocks in words.

1

5

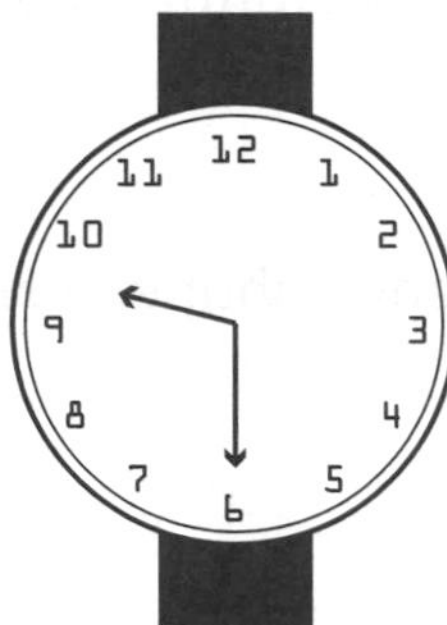

2

6

3

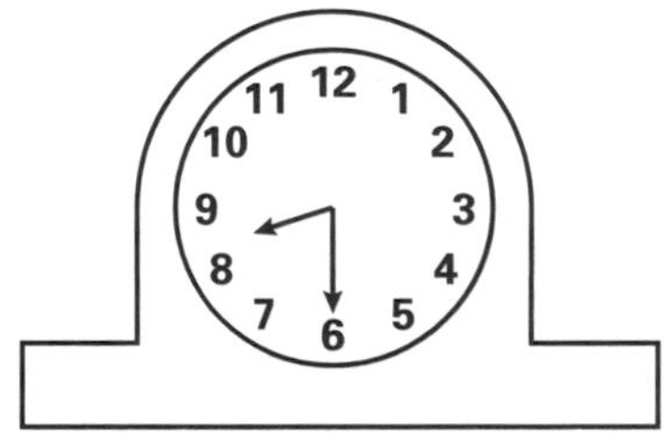

7

4

8

7 3-D work: the extra dimension

CORE

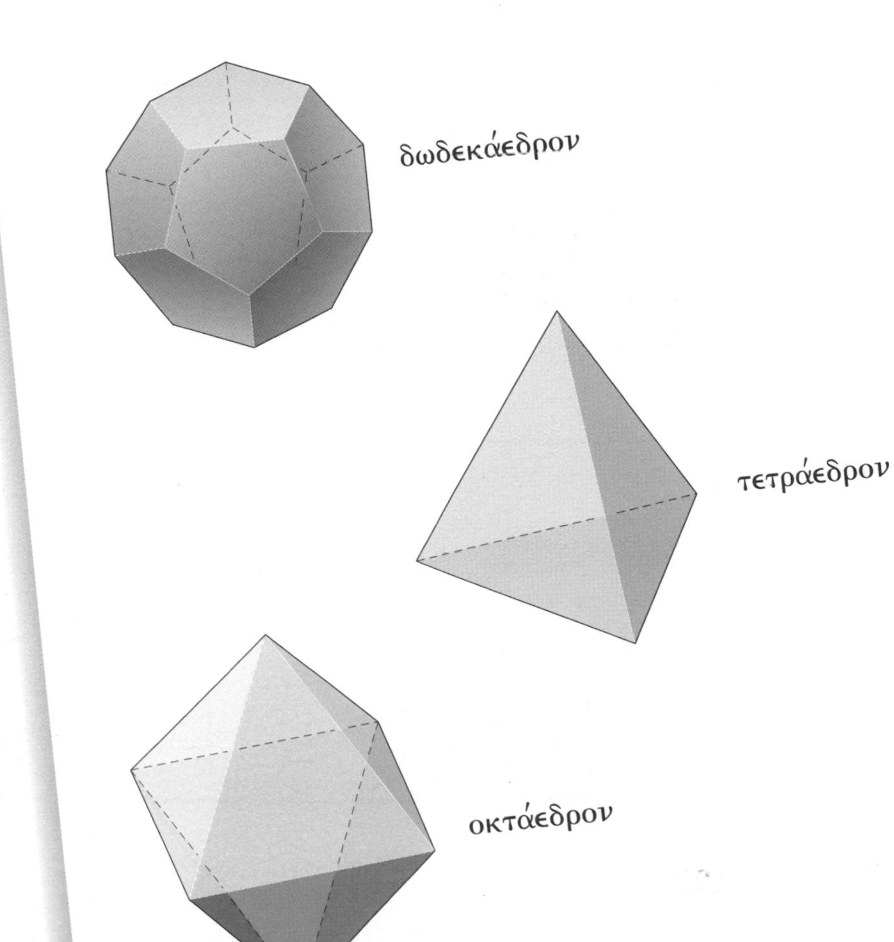

1 Identifying solids

Most picture books have pictures which are in two dimensions (2-D). They are flat on the page, like this one.

Some books work in three dimensions (3-D) by having pop-up pages like this.

Cubes and cuboids are three-dimensional shapes.

Cube A **cube** is a box with square faces.

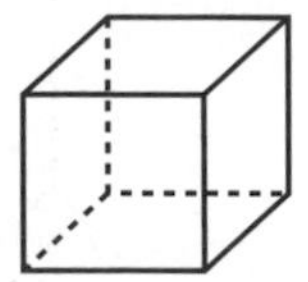

Cuboid A **cuboid** is a box with rectangular faces.

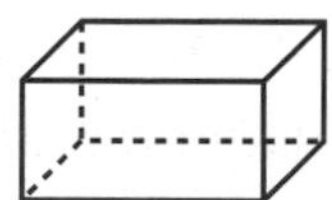

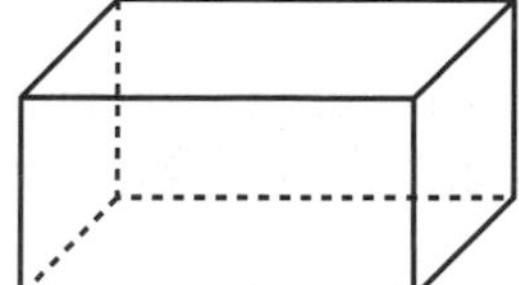

Exercise 7:1

Sort these solids into 2 groups.
Group 1: Cubes.
Group 2: Cuboids.

a

c

e

b

d

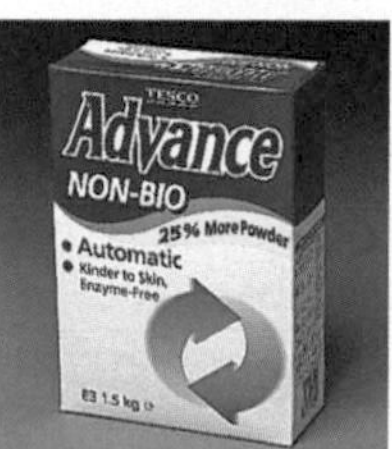

f

Prism A **prism** has the same shape all the way through.

Triangular prism

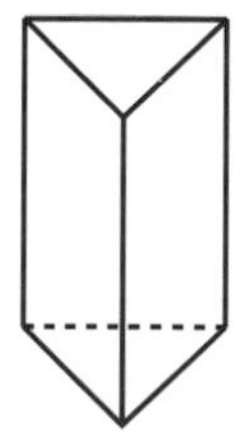

Hexagonal prism

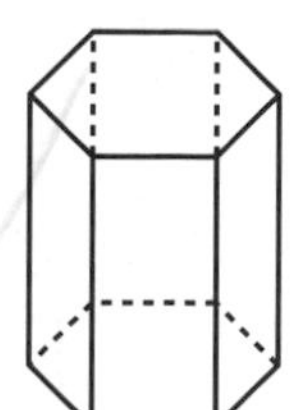

Cylinder

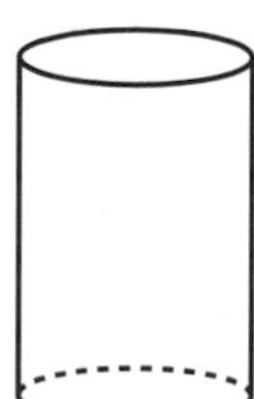

Exercise 7:2

Write the name of the solid for each picture.

a

b

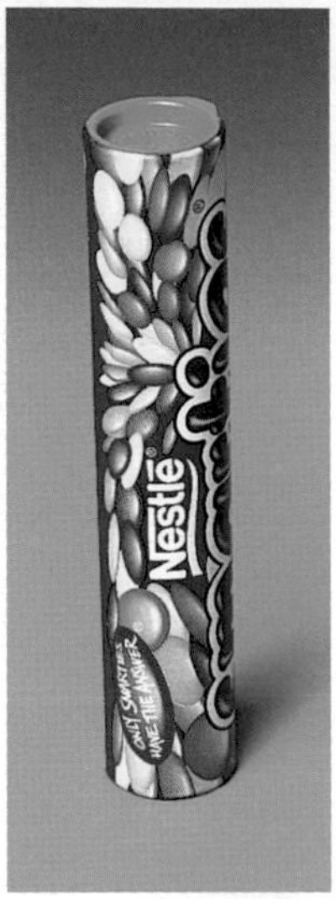

c

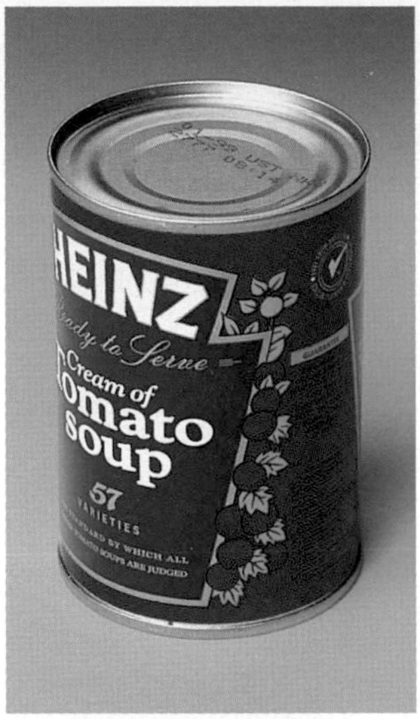

d

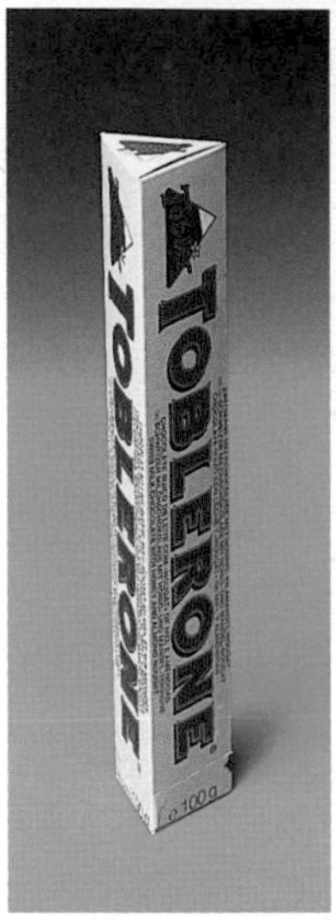

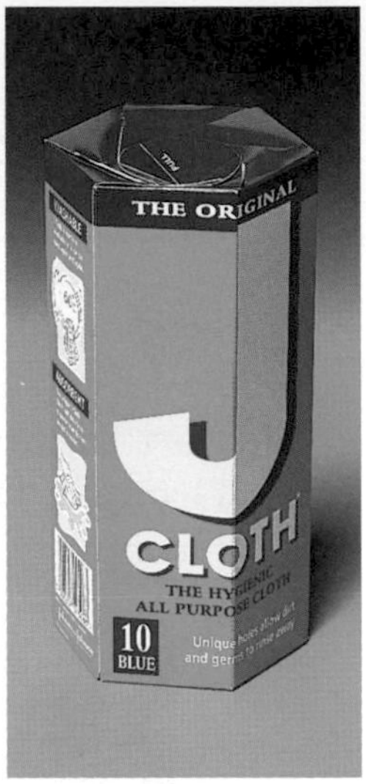

Pyramid

A **pyramid** is a solid with side edges that meet in a point. All of its faces are triangles. The base can be any polygon.

Triangular based pyramid

Square based pyramid

Hexagonal based pyramid

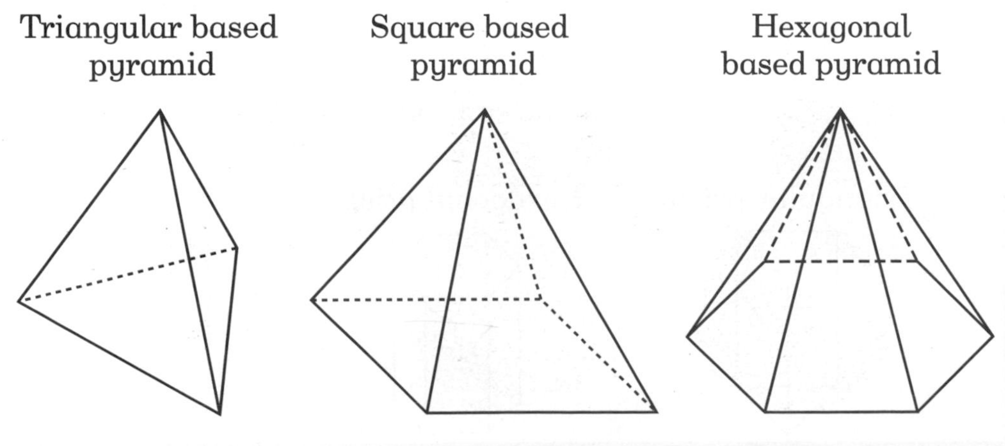

Exercise 7:3

Write the name of the solid for each picture.

a

c

b

d

Cone

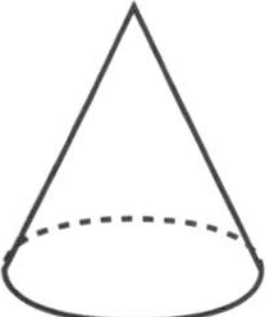

Sphere

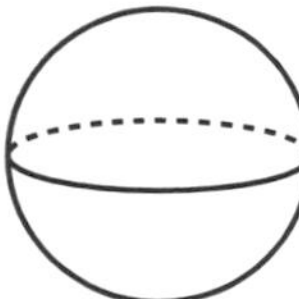

Exercise 7:4

Write the name of the solid for each picture.

a

c

e

b

d

f

G 1, 2 Can you think of any other everyday objects that are interesting solids?

1 Probability scales

The weather map shows more sun symbols than clouds.
A weather presenter says:

'There is a low chance of rain. There is a higher chance of sunshine.'

This is called **probability**.

Probability In maths **probability** means how likely something is to happen.
Probabilities can be shown on a scale with **impossible** at one end and **certain** at the other.

Example Here is a probability scale.

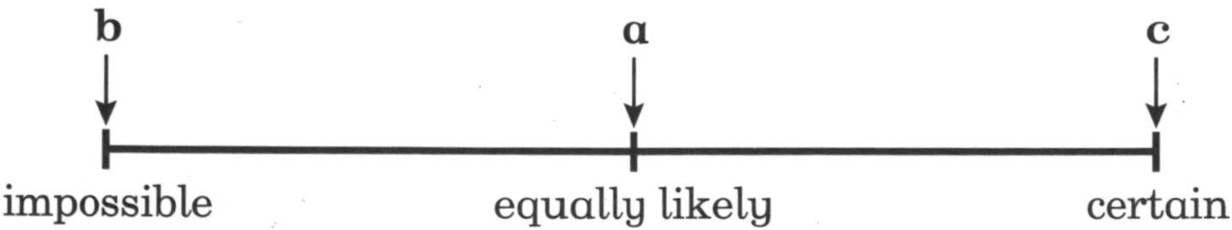

You can see on the scale the probability that:

a a newly born baby will be a girl,

b you will live to be 200,

c it will get light tomorrow.

Exercise 8:1

1 Mark on a probability scale points **a**, **b** and **c** to show how likely you think each one is.

1 **a** You will go home from school today.
b A newly born baby will be a boy.
c You will travel to the moon in one day.

2 **a** A coin thrown in the air will land heads up.
b It will snow in Britain on 23rd July.
c You will talk to someone today.

3 **a** A coin thrown in the air will land tails up.
b You will have a drink today.
c Your teacher will turn into a bookcase.

4 **a** You will travel to Barbados today.
b Someone in your class owns a dog.
c November will follow October.

5 **a** You will get a card for your birthday.
b A dice rolled on a table will land on an even number.
c You will travel to Mars tomorrow.

6 **a** You will sharpen a pencil this year.
b You will buy a tractor next week.
c A dice rolled on a table will land on an odd number.

Some things are **more likely** to happen than others.
I am **more likely** to see a dog than a hippopotamus.

---

7 For each question write down which is **more likely** to happen.

a For breakfast you will have toast.
For breakfast you will have onions.

b You will travel to school by horse.
You will travel to school by bus.

c You will have a holiday in Britain.
You will have a holiday on the moon.

d Tomorrow it will rain for 1 hour.
Tomorrow it will rain for 20 hours.

e You will go to the cinema this week.
You will go to Japan this week.

f Your friend will turn into an alien.
You will get a present for your birthday.

g Your pet is a cat.
Your pet is a lion.

8 For each question write down which is **more likely** to happen.

a You will do 1 hour of homework tonight.
You will do 8 hours of homework tonight.

b You will see a red car today.
You will see a red tractor today.

c You will buy sweets this week.
You will buy a boat this week.

d Your teacher has blue eyes.
Your teacher has red eyes.

e For lunch you will have a sandwich.
For lunch you will have frog's legs.

f Your pet is a dog.
Your pet is a tiger.

g You will travel to school by Concorde.
You will travel to school by bicycle.

H 2

2 How probability works

Hilary is buying a ticket on 'Spin the Wheel'.
Hilary chooses 1 colour.
There are 5 colours.
Hilary has 1 chance out of 5 of winning.

In maths you write:

The probability of Hilary winning is $\frac{1}{5}$.

Exercise 8:2

Philip, Elaine and Alan are playing 'Spin the Wheel'.
There are 5 colours.

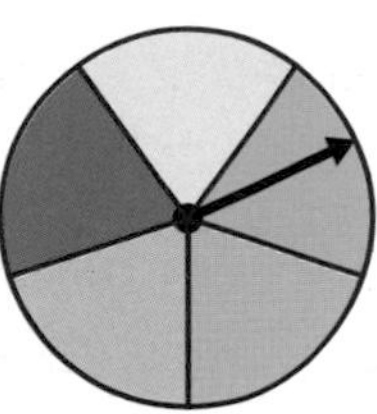

1 **a** Philip buys 1 colour.
Write down the probability that Philip wins.

b Elaine buys 2 colours.
Write down the probability that Elaine wins.

c Alan buys 2 colours.
Write down the probability that Alan wins.

2 Cyril, Richard and Heather are playing 'Spin the Wheel'.
There are 7 colours.

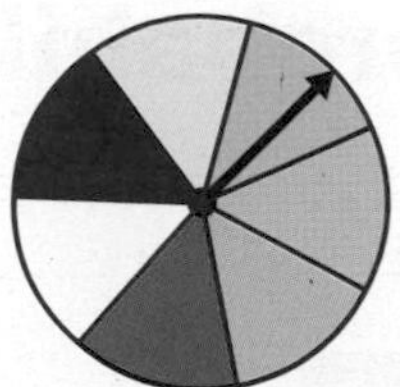

a Richard buys 1 colour.
Write down the probability that Richard wins.

b Cyril buys 2 colours.
Write down the probability that Cyril wins.

 1

c Heather buys 4 colours.
Write down the probability that Heather wins.

Paula is choosing 1 letter from the words KEY MATHS.
There are 8 different letters.
Paula chooses an M.
There is 1 letter M.

The probability that Paula chooses an M is $\frac{1}{8}$.

---

3 **a** Write down the probability that Paula chooses an S.

b Write down the probability that Paula chooses an E.

c Write down the probability that Paula chooses an H.

d Write down the probability that Paula chooses a K.

e Write down what you notice about your answers.

This is a 6-sided dice.
The numbers on it are 1, 2, 3, 4, 5 and 6.
Each number has the same chance of being rolled.
In maths, you say that each number has an **equal chance** of being rolled.

Brian is rolling a 6-sided dice.
There are **6** different numbers.
Brian rolls a three.
There is **1** of these.
The probability that Brian rolls a three is $\frac{1}{6}$.

 3

4 **a** Write down the numbers on a 6-sided dice.

b Write down the probability of rolling a two.

c Write down the probability of rolling a one.

d Write down the probability of rolling a five.

e Write down the probability of rolling a four.

f Write down the probability of rolling a six.

g Write down what you notice about your answers.

Example Write down the **probability** of rolling an **even number** on a 6-sided dice.

There are **3** even numbers on a **6**-sided dice.

The probability of rolling an even number is $\frac{3}{6}$.

You can use a 6-sided dice to help you with Questions **5**, **6** and **7**.

5 **a** Write down the odd numbers on a 6-sided dice.

b How many odd numbers are there?

c Write down the probability of rolling an odd number on a 6-sided dice.

6 **a** Write down the numbers less than three on a 6-sided dice.

b How many numbers are less than three?

c Write down the probability of rolling a number which is less than three on a 6-sided dice.

7 **a** Write down the numbers bigger than two on a 6-sided dice.

b How many numbers are bigger than two?

c Write down the probability of rolling a number bigger than two on a 6-sided dice.

Example This is a 4-sided dice.
What is the probability of rolling a 3?
The numbers on the dice are 1, 2, 3 and 4.
There are **4** different numbers.
Only **1** of the numbers is a 3.
The probability of rolling a 3 is $\frac{1}{4}$.

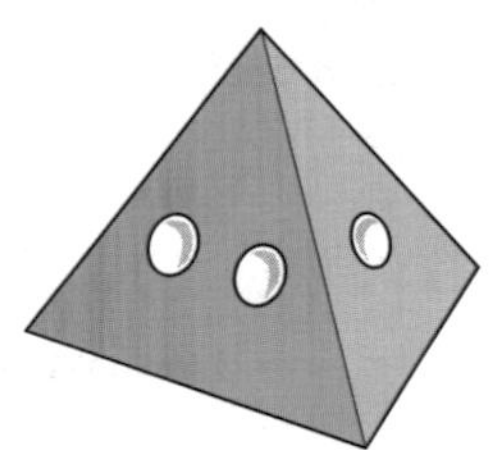

You can use a 4-sided dice to help you.

8 **a** Write down the numbers on a 4-sided dice.

b Write down the probability of rolling a two.

c Write down the probability of rolling a one.

d Write down the probability of rolling a four.

e Write down what you notice about your answers.

Example Hazel puts these **7** counters into a bag.

She asks Roger to choose a counter without looking.
The probability that he chooses a red counter is $\frac{3}{7}$.
The probability that he chooses a yellow is $\frac{4}{7}$.

Exercise 8:3

You can use counters to help you with this exercise.

1 Ruth puts 5 counters in a bag.
2 counters are red and 3 counters are yellow.
She asks Patrick to choose a counter without looking.

a Write down the probability that he chooses red.

b Write down the probability that he chooses yellow.

2 Nigel puts 6 counters in a bag.
2 counters are red and 4 counters are yellow.
He asks David to choose a counter without looking.

a Write down the probability that he chooses red.

b Write down the probability that he chooses yellow.

3 Lesley puts 11 counters in a bag.
10 counters are red and 1 counter is yellow.
She asks John to choose a counter without looking.

a Write down the probability that he chooses red.

b Write down the probability that he chooses yellow.

4 Linda has a bag of sweets. There are 11 soft sweets and 10 hard sweets. She offers the bag to Christine.

a Write down the probability that she chooses a hard sweet.

b Write down the probability that she chooses a soft sweet.

A pack of playing cards has 4 suits.

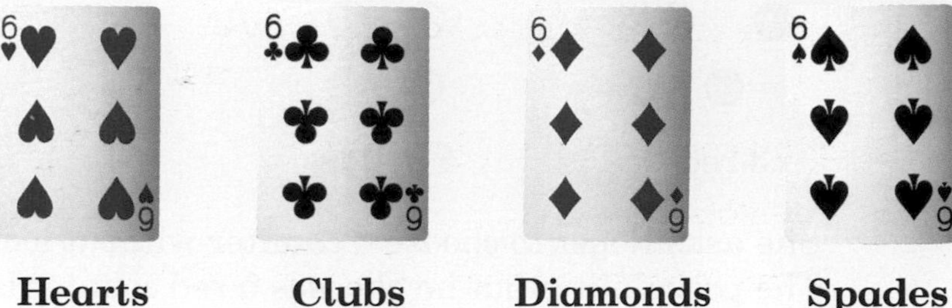

Hearts **Clubs** **Diamonds** **Spades**

Example Megan has a pack of cards. There are **52** cards.
She counts the diamonds. There are **13** diamonds.
She shuffles the cards.
Jon chooses a card.
The probability he chooses a diamond is $\frac{13}{52}$.

Exercise 8:4

H 5 You can use a pack of cards to help you with this exercise.

1 **a** How many cards are there in a pack?

b How many cards are black?

c What is the probability of choosing a black card?

2 **a** How many cards are there in a pack?

b How many cards are red?

c What is the probability of choosing a red card?

3 **a** How many cards are there in a pack?

b How many cards are hearts?

c What is the probability of choosing a heart?

4 **a** How many cards are there in a pack?

b How many cards are spades?

c What is the probability of choosing a spade?

5 **a** How many cards are there in a pack?

b How many fives are there?

c What is the probability of choosing a five?

6 **a** How many cards are there in a pack?

b How many aces are there?

c What is the probability of choosing an ace?

7 **a** How many cards are there in a pack?

b How many black sixes are there?

c What is the probability of choosing a black six?

Higher or Lower

This is a card game for 2 players.
You need one suit of cards.
Choose hearts, diamonds, spades or clubs.

Rules

1. Shuffle the cards.
2. Deal the 13 cards face down in a line.
3. Turn over the first card on the left.

4. The first player decides if the next card is going to be **higher** or **lower**.
5. They turn over the next card.
   If the player is right they score 1 point.
6. Now it is the second player's turn.
7. Keep playing until all the cards are turned over.
8. The winner is the player with the most points.

Play the game 3 times or for 5 minutes.

3 Experiments and games

Probability can help you when you are playing games.

Odds and Evens

 2

This game is for 2 players.
You need a 6-sided dice.

One player is **Evens** and the other player is **Odds**.
Take it in turns to throw the dice.

Evens throws first. If the dice shows an even number, then score 1 point. If it is odd, score no points.
Odds goes next and only scores a point if the dice shows an odd number.

Tally your score in a table like this:

Game	Evens	Odds	Winner
1	𝍸	𝍷𝍷𝍷	Evens
2	𝍷𝍷𝍷𝍷	𝍸	Odds

The first player to score 5 points wins the game.

Play the game 6 times.

1 How many games did **Evens** win?

2 How many games did **Odds** win?

3 Was this what you expected?
Write a sentence to say why.

4 What is the probability of scoring an even number on any throw?

5 What is the probability of scoring an odd number on any throw?

Under and Over 4

 3

This game is for 2 players.
You need a 6-sided dice.

One player chooses numbers under 4, the other player chooses numbers over 4.
Take it in turns to throw the dice.

If **Under 4** throws a number under 4, they score 1 point.
If **Over 4** throws a number over 4, they score 1 point.

Tally your score in a table like this:

Game	Player A Under 4	Player B Over 4	Winner
1	卌 卌 ||	卌 |||	Under 4
2			
3			
4			
5			
6			

Each player throws the dice 10 times.
The winner is the player with the highest tally.

You can play the game 6 times.

1 Do you think this is a fair game?

2 How many times did **Under 4** win?

3 How many times did **Over 4** win?

4 How could you change the rules of the game to make it fair?

Make a game

A game is fair if everyone has the same chance of winning.

This is a 12-sided dice.

Make a game using a 12-sided dice.

Make your game fair.

Write the rules of your game so that someone else can play it.

Example Aisha is planning a probability experiment.

She has 4 red counters and 2 green counters in a bag.

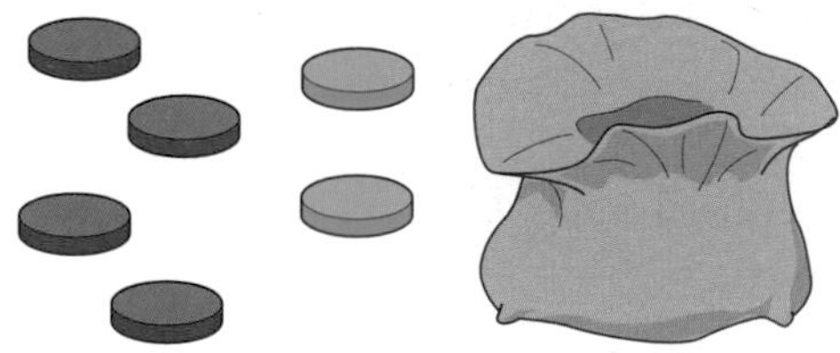

Aisha is going to pick a counter without looking.
She records the colour in a tally-table.
She then puts the counter back.

She does it 12 times.

She thinks that she will get 8 red counters and 4 green counters because the number of red counters is double the number of green counters.

An experiment does not always give you the result you expect.

Red	Green
卌 ||||	|||

She gets 9 red counters and 3 green counters.

Exercise 8:5

1 Do a probability experiment with 4 red counters and 2 green counters and a bag.
You can do this experiment with a friend.

a Choose a counter.
Record the colour in a tally-table.
Put your counter back.
Do this 6 times.
What did you expect to get?
What did you get?

b Choose a counter 12 times.
Record the colour in a tally-table.
Put your counter back.
What did you expect to get?
What did you get?

c Choose a counter 18 times.
Record the colour in a tally-table.
Put your counter back.
What did you expect to get?
What did you get?

d Choose a counter 24 times.
Record the colour in a tally-table.
Put your counter back.
What did you expect to get?
What did you get?

2 Choose different numbers of counters.
You could choose 4 red counters and 4 green counters.
Do the experiment again.

1 For each question below draw a probability scale. Mark on it points **a**, **b** and **c** to show how likely you think each one is.

a You will not sleep for a month.

b You will wake up today.

c A coin thrown up in the air will land heads down.

2 Which is more likely to happen?

a You will have sausage and chips for tea.

b You will have cornflakes for tea.

3 Which is more likely to happen?

a You will buy a motorbike this week.

b You will buy a packet of sweets this week.

4 Jonathan is playing 'Spin the Wheel'. There are 3 colours.

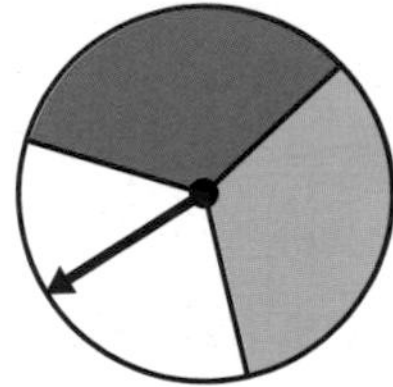

Jonathan chooses 1 colour. Write down the probability that Jonathan wins.

5 Ruth and Simon are playing 'Spin the Wheel'. There are 7 colours.

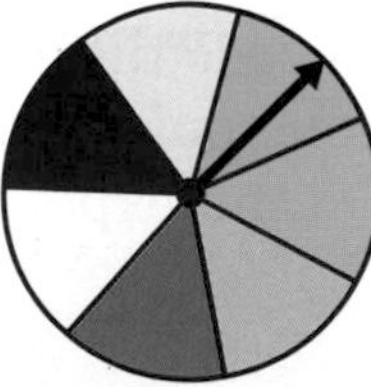

a Ruth buys 1 colour. Write down the probability that Ruth wins.

b Simon buys 4 colours. Write down the probability that Simon wins.

6 There are 7 letters in the word TUESDAY. Dave chooses a letter. Write down the probability that he chooses an E.

7 **a** Write down the numbers on a 6-sided dice.

b Write down the probability of rolling a two.

8 **a** Write down the numbers bigger than four on a 6-sided dice.

b How many numbers bigger than four are there?

c Write down the probability of rolling a number bigger than four on a 6-sided dice.

9 **a** Write down the numbers less than four on a 6-sided dice.

b How many numbers less than four are there?

c Write down the probability of rolling a number less than four.

10 Geoff puts 4 counters in a bag.
3 are white and 1 is blue.
Geoff asks Joy to choose a counter without looking.

a Write down the probability that she chooses a white counter.

b Write down the probability that she chooses a blue counter.

11 Ashok puts 7 counters in a bag.
2 are red and 5 are yellow.
Sara chooses a counter without looking.

a Write down the probability that she chooses a red counter.

b Write down the probability that she chooses a yellow counter.

12 **a** How many cards are there in a pack of playing cards?

b How many are black?

c What is the probability of choosing a black card?

13 **a** How many cards are there in a pack?

b How many are clubs?

c What is the probability of choosing a club?

14 **a** How many cards are there in a pack?

b How many nines are there?

c What is the probability of choosing a nine?

1 Take it away!

Ted's Take-away

Exercise 9:1

1 **a** How much is a carton of chips?

b How much is a burger?

c How much is a pizza?

d How much is a sandwich?

e How much is a hot dog?

2 This tray has four pizzas on it.

Write down what is on each tray.

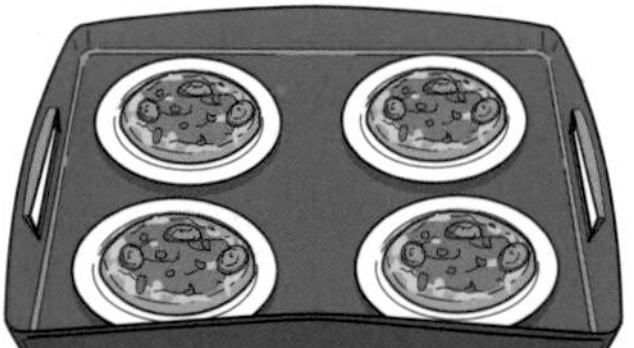

a

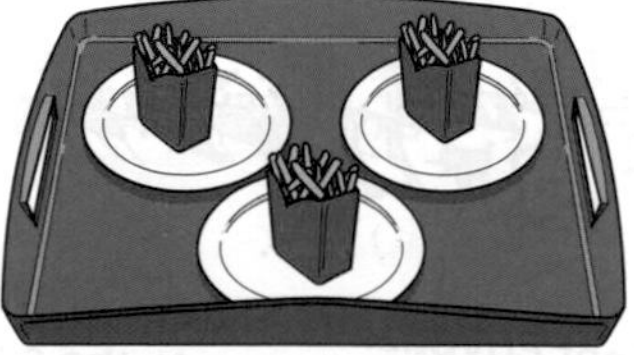

c

b

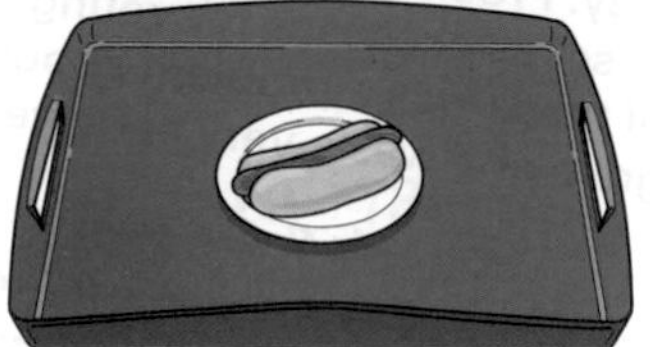

d

Example How much you pay depends on how much you buy.

The cost of 1 hot dog is 15 p × 1
The cost of 2 hot dogs is 15 p × 2
The cost of 3 hot dogs is 15 p × 3
The cost of 4 hot dogs is 15 p × 4
The cost of 5 hot dogs is 15 p × 5

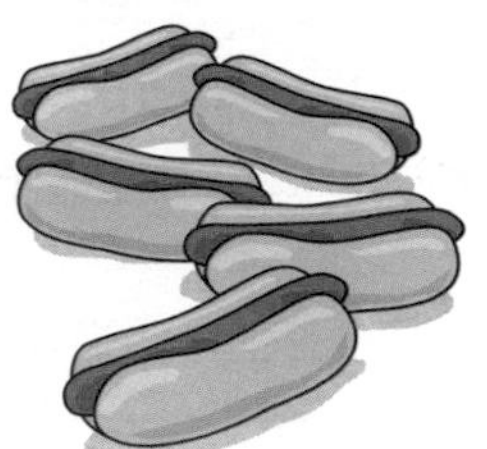

You can see that:

The total cost = 15 p × the number of hot dogs.

W 1

3 Fill in the worksheet.

a The cost of 1 carton of chips is 25 p × 1
The cost of 2 cartons of chips is 25 p × ☐
The cost of 3 cartons of chips is ☐ × 3
The cost of ☐ cartons of chips is 25 p × 4
The cost of 5 cartons of chips is ☐ × ☐

The total cost = ☐ p × the number of cartons of chips.

b The cost of 1 pizza is 20 p × 1
The cost of 2 pizzas is ☐ × 2
The cost of ☐ pizzas is 20 p × ☐
The cost of ☐ pizzas is ☐ × ☐
The cost of ☐ pizzas is ☐ × ☐

The total cost = ☐ p × the number of pizzas.

c The cost of 1 burger is 30 p × 1
The cost of 2 burgers is ☐ × 2
The cost of ☐ burgers is 30 p × ☐
The cost of ☐ burgers is ☐ × ☐
The cost of ☐ burgers is ☐ × ☐

H 1 The total cost = ☐ p × the number of burgers.

Example Toby and Sanjay go to a restaurant.

They leave a tip of 50 p.

They pay the cost of the meal + 50 p.

W 2 **4** Write these tips in the same way.

a Helen and Kiran leave a tip of 20 p.

They pay the cost of the meal + ☐

b Philip and Sue leave a tip of 60 p.

They pay the cost of the meal + ☐

Example Steve goes to a restaurant on Monday.

On Monday, every meal has 40 p off.

He pays the cost of the meal − 40 p.

W 2 **5** Write these tips in the same way.

a On Tuesday, every meal has 50 p off.

Steve pays the cost of the meal − ☐

b On Wednesday, every meal has 45 p off.

Steve pays the cost of the meal − ☐

2 Taking your order

Molly cannot write down
Mike's order fast enough.
She writes it down in code.

This order can be written in words:

one ***b***urger and one ***c***arton of chips

This order can be written in code:

$1b + 1c$

Exercise 9:2

Write these orders in words and in code.

1 **a**

b

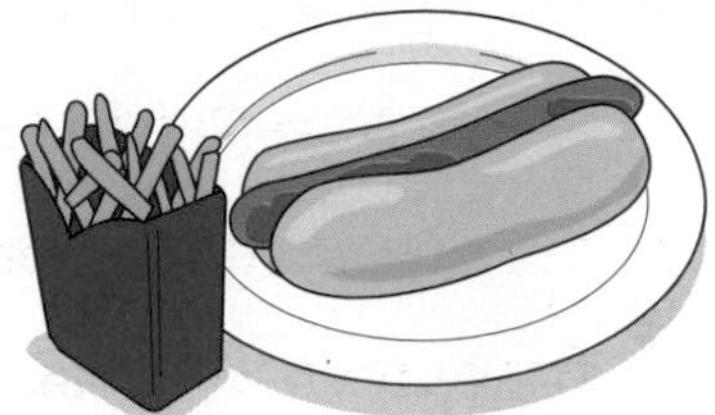

c

e

d

f

Example This order can be written in words:

*s*ausage and *s*ausage

or two sausages

This order can be written in code:

$1s + 1s$ or $2s$

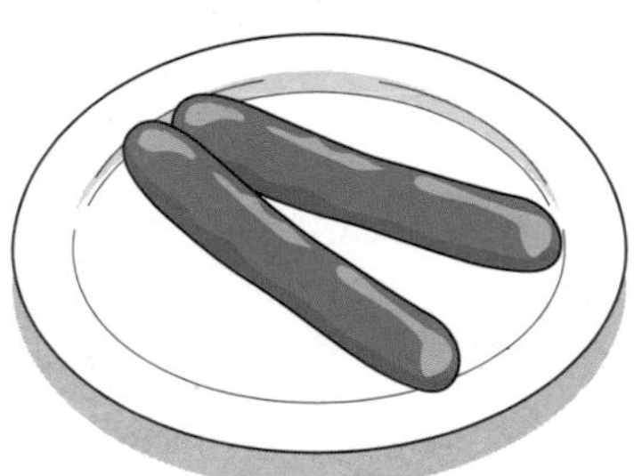

---

2 Write these orders in words and in code.

a

c

b

d

e

f

Example This order can be written in words:

two sausages and one carton of chips

This order can be written in code:

$2s + 1c$

---

3 Write these orders in words and in code.

a

d

b

e

c

f

H 5 **2** Write each of these in the same way.

a

$a + a + t + t + t = \square\, a + \square\, t$

b

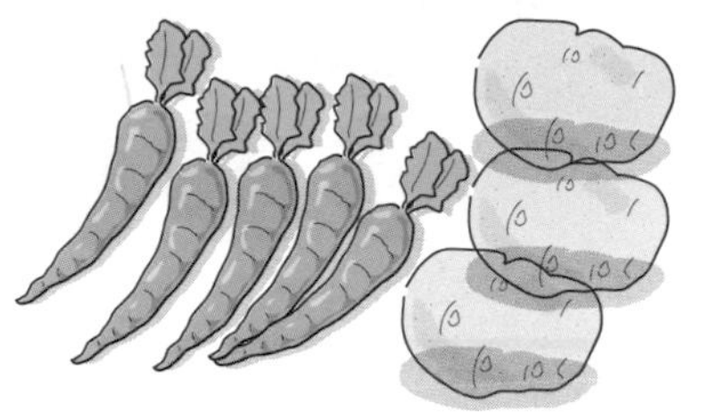

$c + c + c + c + c$

$+ p + p + p = \square\, c + \square\, p$

c

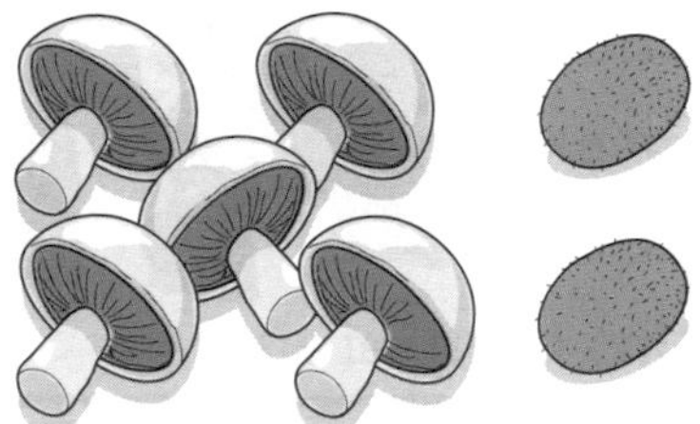

$m + m + m + m + m$

$+ k + k = \square \ldots + \square \ldots$

d

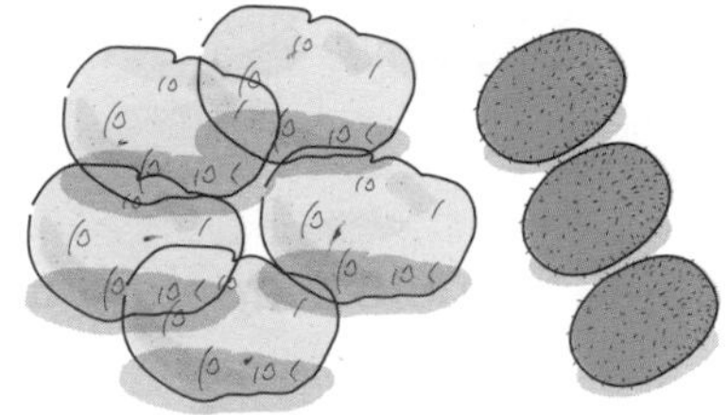

H 6

G 1, 2

$p + p + p + p + p$

$+ k + k + k = \square \ldots + \square \ldots$

e

f

g

h

Example You can collect terms like this too:

$3t + 5t = 8t$

3 Collect these terms.

a $3p + 4p =$

b $2t + 4t =$

c $3s + 5s =$

d $5a + 6a =$

e $5b + 4b =$

f $6c + 7c -$

g $6m + 8m =$

h $7l + 4l =$

Example To collect terms you add the letters that are the same.

$2s + 3s + 2t + 4t = 5s + 6t$

4 Collect these terms.

a $3d + d + e + 3e = \square\ d + \square\ e$

b $5a + 2a + 2b + 3b = \square\ a + \square\ b$

c $2p + 4p + q + 3q = \square \ldots + \square \ldots$

d $3g + 4g + 2h + h = \square \ldots + \square \ldots$

e $4y + y + 3z + 2z = \square \ldots + \square \ldots$

5 Collect these terms.

a $2d + 4d + 3e = \square\ d + \square\ e$

b $3r + 2r + 4s = \square \ldots + \square \ldots$

c $2a + 3b + 2b = \square \ldots + \square \ldots$

d $3c + 5c + 4m = \square \ldots + \square \ldots$

e $2f + 4f + 7g = \square \ldots + \square \ldots$

5 Codes

Using letters for numbers is called **algebra**.

Example Words can be made using code.

s	p	o	r	t
19	16	15	18	20

a	p	p	l	e
1	16	16	12	5

Exercise 9:5

 7

1 Find the codes for these words.

a book
b house
c clock
d cow
e rabbit
f onion
g pizza
h burger
i tomato
j mathematics

H 7

2 Find the codes for these sentences.

a Make a cake.
b Tell the time.
c Maths is fun.
d I am twelve years old.
e Can you keep a secret?

3 Find the words from these codes.

a

3	15	4	5	19

b

13	5	19	19	1	7	5

c

19	17	21	1	18	5

d

3	9	18	3	12	5

e

1	12	7	5	2	18	1

4 Find the sentences from these codes.

a

8	1	16	16	25	2	9	18	20	8	4	1	25

b

3	21	19	20	1	18	4	6	15	18	20	5	1

c

23	5	23	1	14	20	9	3	5	3	18	5	1	13

5 Find the words from these codes.

a

1	14	9	13	1	12

b

8	1	13	19	20	5	18

c

18	1	2	2	9	20

d

8	9	16	16	15	16	15	20	1	13	21	19

e

7	9	18	1	6	6	5

6 Find the sentences from these codes.

a

16	1	18	20	25	20	9	13	5

b

3	8	18	9	19	20	13	1	19	3	1	18	4

7 Use the code to write some messages of your own.
Send a note in code to a friend.

1 Write down what is on each tray.

a

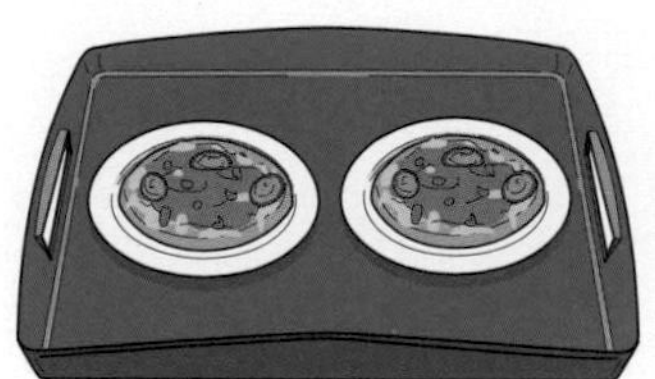

c

b

d

2 **a** Nathan and Fiona leave a tip of 40 p.

They pay the cost of the meal + ☐

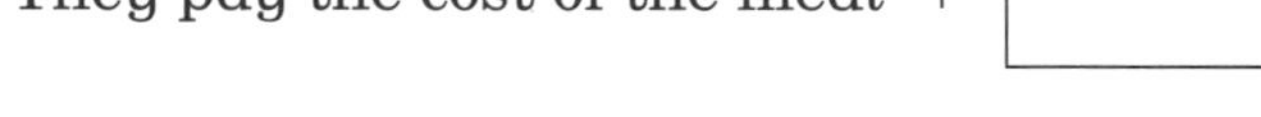

b Stuart and Kathy leave a tip of 30 p.

They pay the cost of the meal + ☐

c Nick and Hilary leave a tip of 65 p.

They pay .... +

3 Collect the terms for these pictures.

a

c

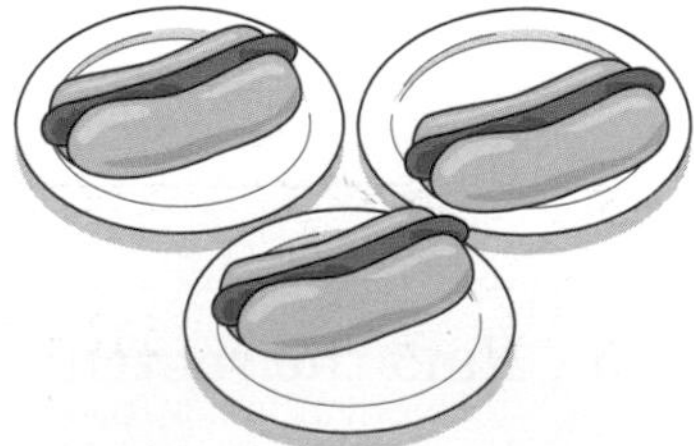

b

d

4 Write these orders in words and in code.

a

b

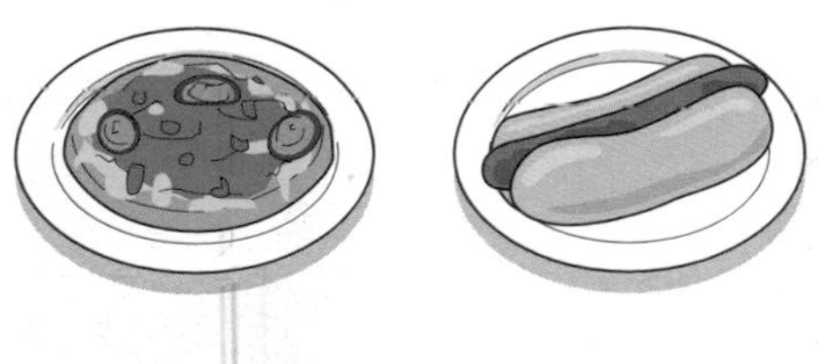

c

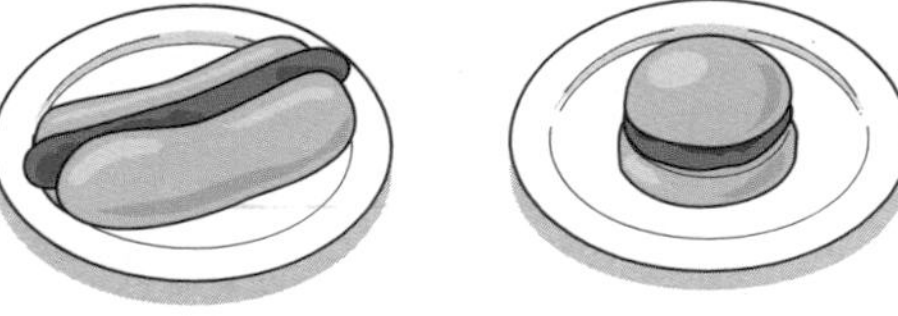

5 Collect these terms.

a $2b + 3b =$

b $6a + 4a =$

c $8s + 9s =$

d $3t + 5t =$

6 Collect these terms.

a $3d + 4e + 2d + 2e = \square \ldots + \square \ldots$

b $2s + 4t + 4s + 4t = \square \ldots + \square \ldots$

c $3x + 4x + 2y + 5y = \square \ldots + \square \ldots$

d $8a + 1b + 3a + 5b = \square \ldots + \square \ldots$

Multiplication 2

Copy these into your book.
Work out the answers.

Exercise 1	Exercise 2	Exercise 3	Exercise 4
1 3 × 10	**1** 5 × 100	**1** 5 × 10	**1** 8 × 10
2 5 × 10	**2** 1 × 10	**2** 15 × 10	**2** 7 × 10
3 7 × 100	**3** 2 × 100	**3** 3 × 10	**3** 9 × 10
4 6 × 10	**4** 4 × 10	**4** 13 × 10	**4** 14 × 10
5 9 × 10	**5** 9 × 10	**5** 7 × 10	**5** 12 × 100
6 3 × 100	**6** 3 × 10	**6** 8 × 10	**6** 41 × 10
7 4 × 100	**7** 7 × 100	**7** 6 × 10	**7** 17 × 100
8 4 × 10	**8** 2 × 10	**8** 2 × 100	**8** 32 × 10
9 8 × 10	**9** 6 × 100	**9** 2 × 10	**9** 35 × 10
10 6 × 100	**10** 6 × 10	**10** 9 × 10	**10** 32 × 10
11 2 × 10	**11** 5 × 100	**11** 3 × 10	**11** 21 × 10
12 10 × 10	**12** 7 × 10	**12** 5 × 10	**12** 16 × 10
13 7 × 10	**13** 3 × 100	**13** 12 × 10	**13** 43 × 10
14 9 × 100	**14** 8 × 10	**14** 4 × 10	**14** 24 × 10
15 10 × 100	**15** 4 × 100	**15** 18 × 10	**15** 47 × 10
16 2 × 100	**16** 8 × 100	**16** 13 × 10	**16** 51 × 10

10 Angles

The Earth's orbit

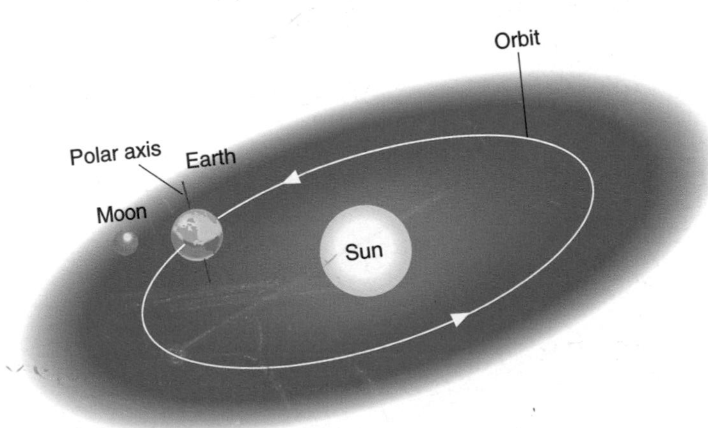

The Earth takes one year, or about $365\frac{1}{4}$ days, to go round the Sun. The $\frac{1}{4}$s add together and every four years make a leap year.

Long ago people thought the Sun went round the Earth. They thought it took 360 days to go round. 360 degrees in a circle probably came from this – one degree for each day.

1 Introducing angles

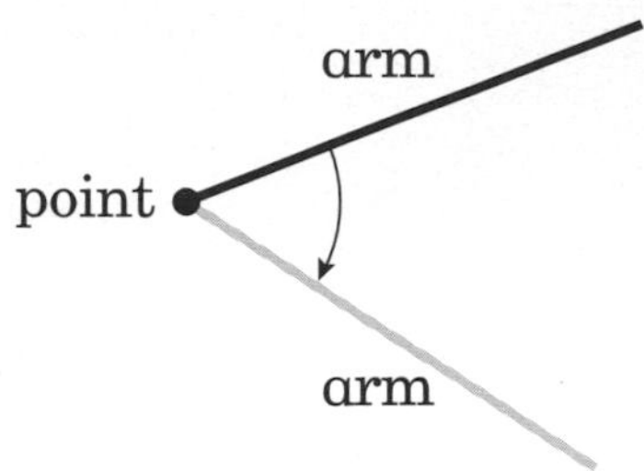

An **angle** has 2 arms which meet at a point. An angle measures an amount of turn.

Exercise 10:1

H 1 **1** Trace these angles carefully.

a

c

b

d

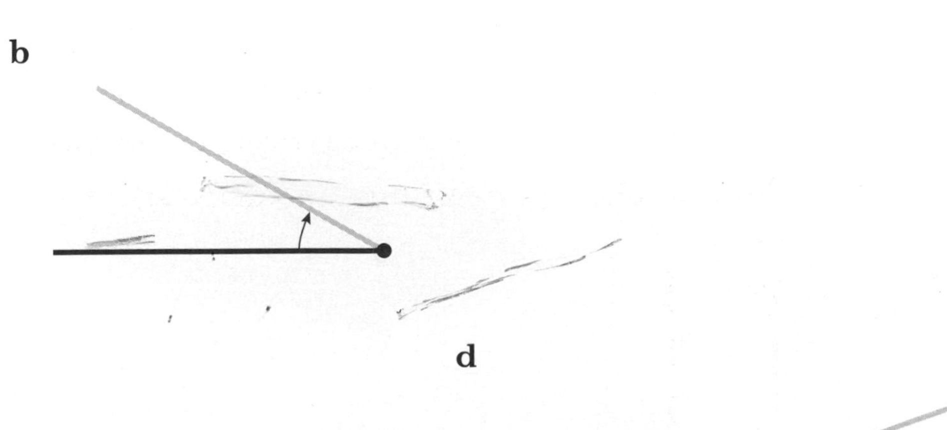

2 Stand your book on your desk.
Look down at it from above.
Try and make the cover look like the angles you have drawn in Question 1.

a

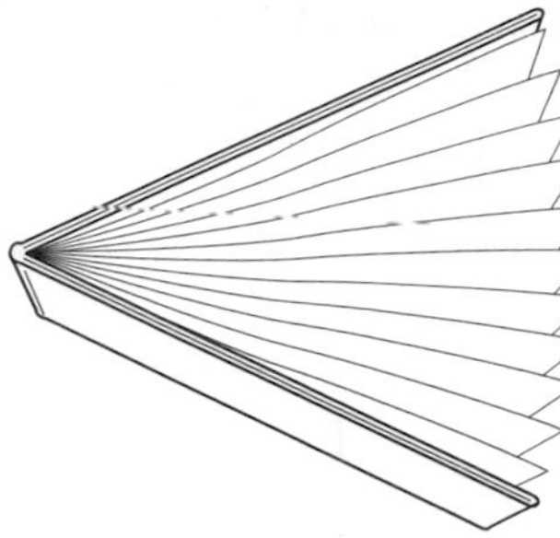

c

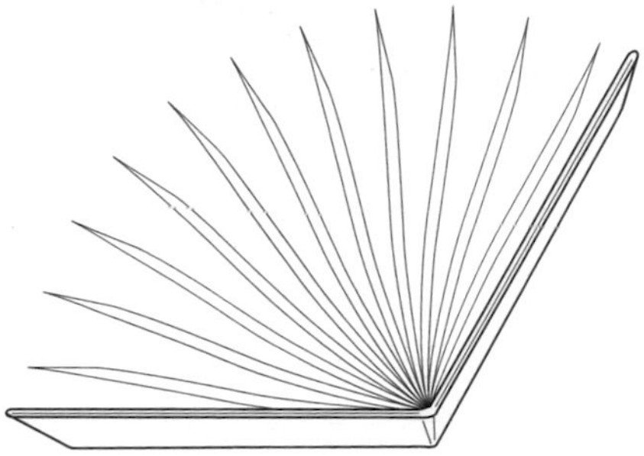

b

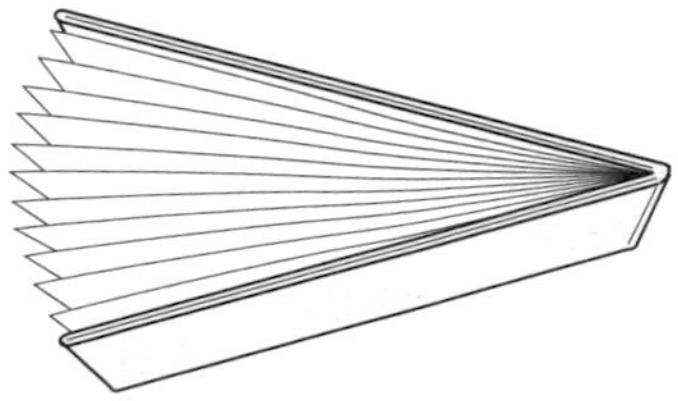

d

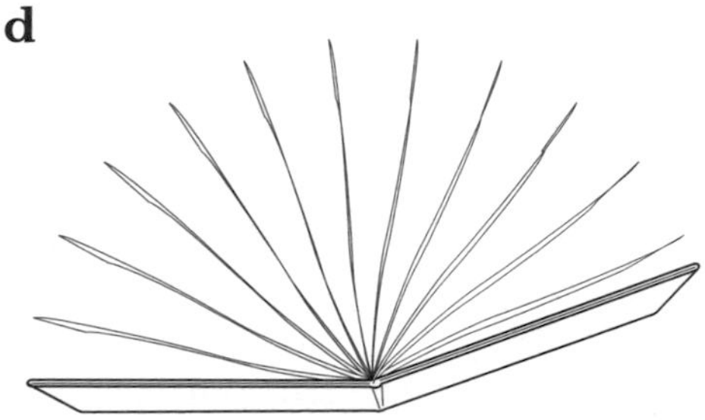

Right angle $\frac{1}{4}$ turn is called a **right angle**.

The corners of squares and rectangles are right angles.

A right angle is often shown like this:

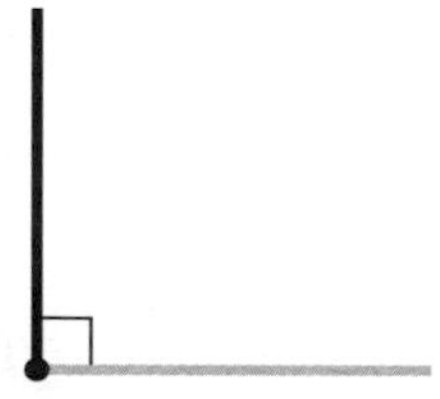

Right angles can be found all around us.

Exercise 10:2

1 Write down 6 things with right angle shaped corners.

2 Trace this right angle very carefully.
You will need it later.

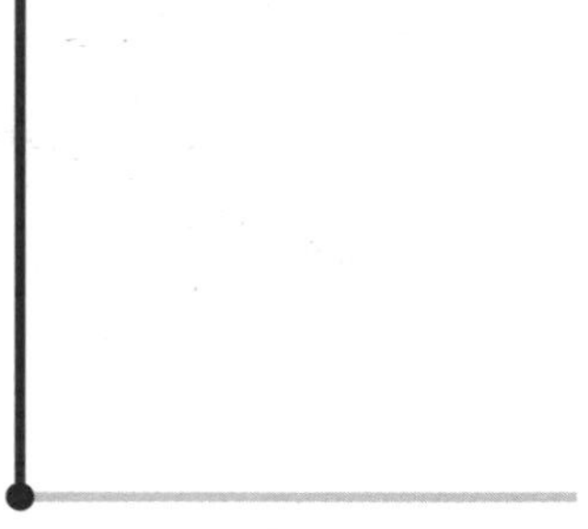

Example Helen uses *her* tracing paper right angle to check if *these* angles are right angles.
She puts the tracing over the angle, with the dark arm on top of the dark arm and the point on top of the point.
She checks if the other arms fit on top of each other.

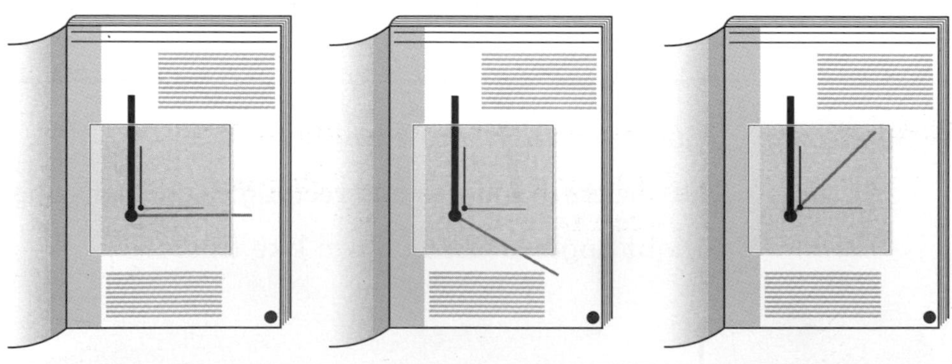

This **is** a right angle.	This **is not** a right angle. It is bigger than a right angle.	This **is not** a right angle. It is smaller than a right angle.

Exercise 10:3

1 Use *your* tracing paper right angle to check if *these* angles are right angles.
Write down either:

This is a right angle.
or This is bigger than a right angle.
or This is smaller than a right angle.

a

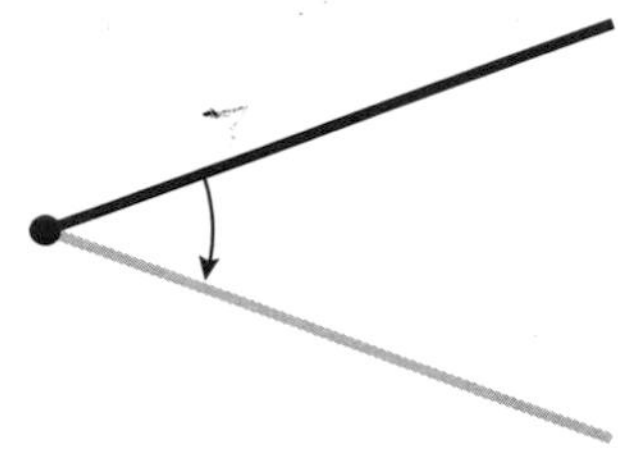

c

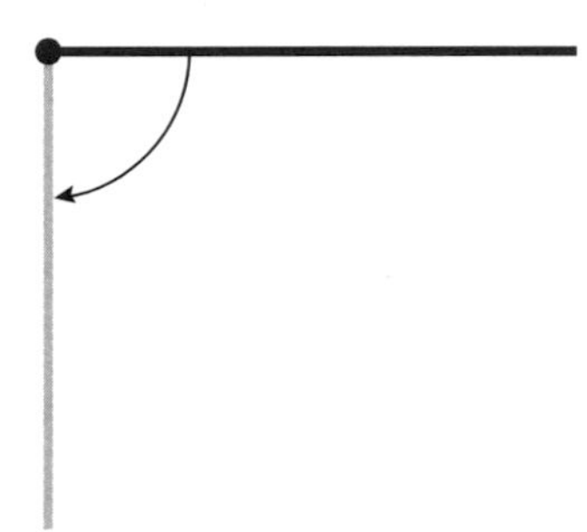

b

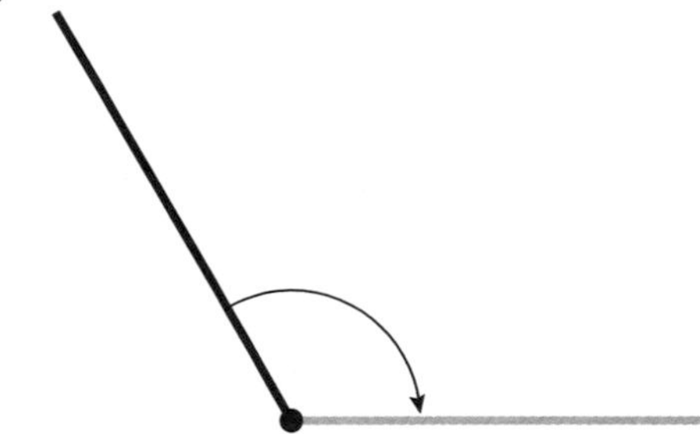

d

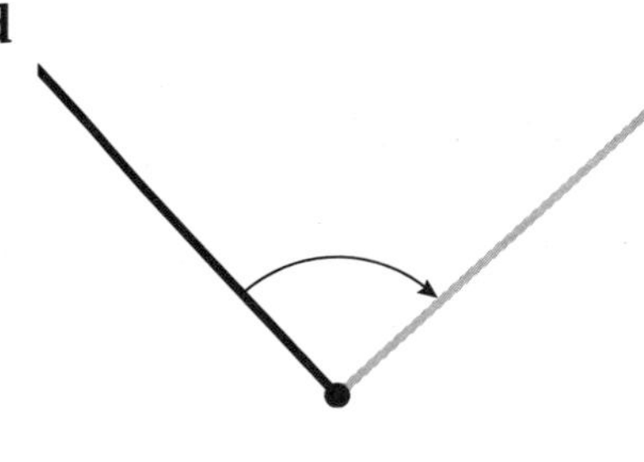

Degrees

You use **degrees** to measure angles.

A right angle measures 90 degrees.
This can be written as 90°.

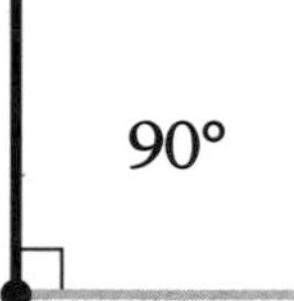

Acute

An angle that is smaller than 90° is called an **acute** angle.

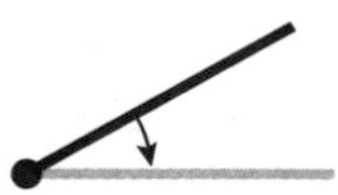

Obtuse An angle that is bigger than 90° (but smaller than 180°) is called an **obtuse** angle.

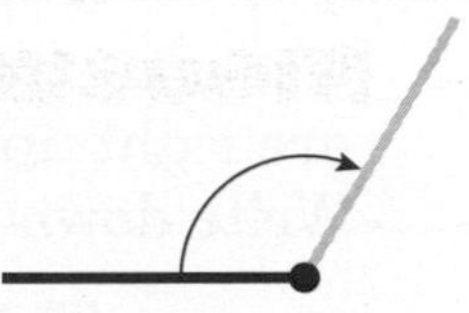

Example Craig is checking if his angles are acute or obtuse. He uses *his* right angle.

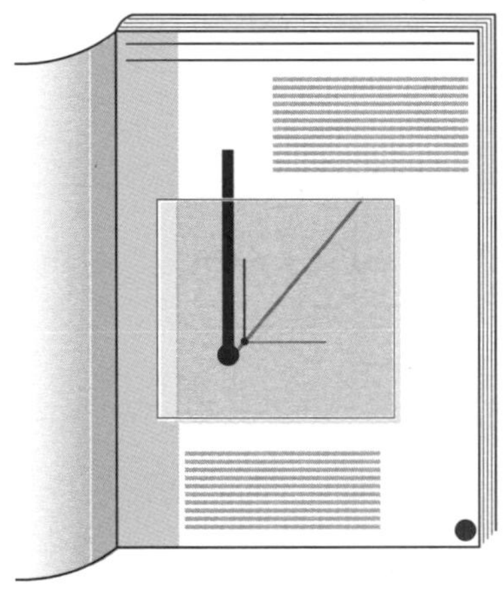

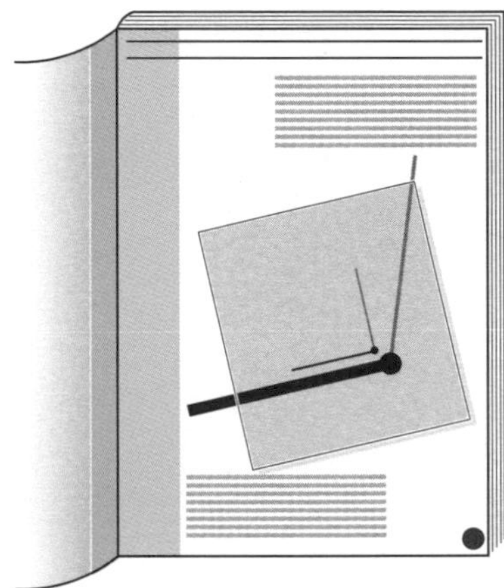

This angle is smaller than 90°. It is an **acute** angle.

This angle is bigger than 90°. It is an **obtuse** angle.

 3

2 Use *your* right angle to check if *these* angles are acute angles or obtuse angles.
Write down either:

This is an acute angle.
or This is an obtuse angle.

a

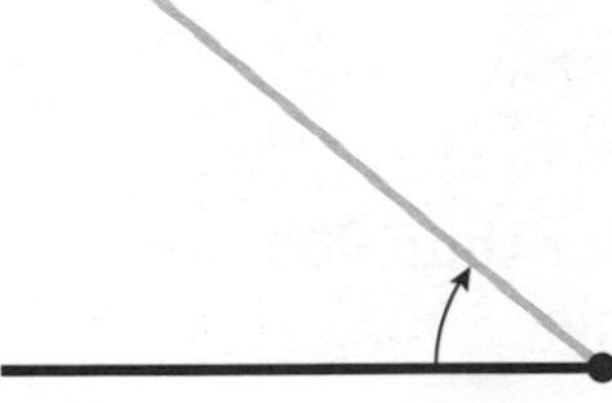

b

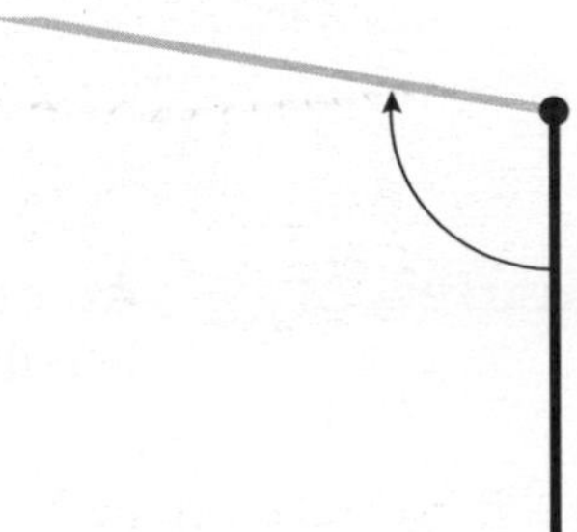

2 Using a protractor or angle measurer

Measuring angles

You use a protractor or angle measurer for **measuring angles**.
A protractor can be in the shape of either a half turn or a full turn.

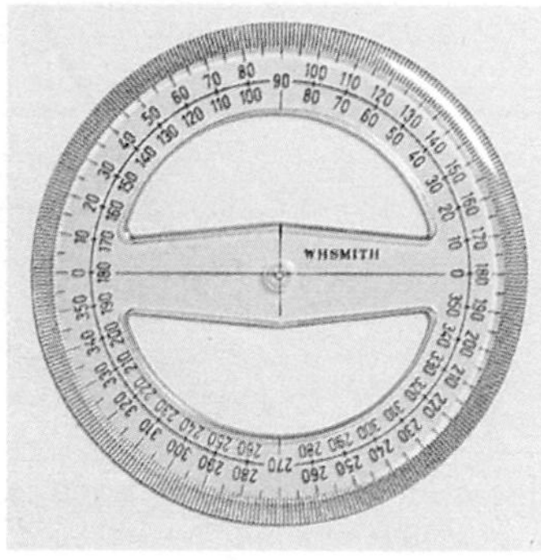

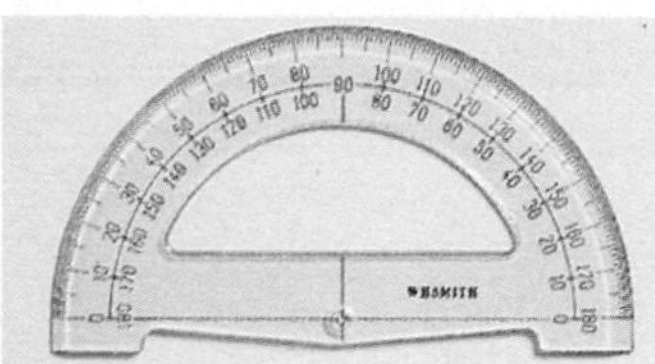

Example

Richard is measuring an angle.
He puts the 0° line on the dark arm of the angle he is measuring.
He counts round in tens from 0°.
He always reads the clockwise scale.

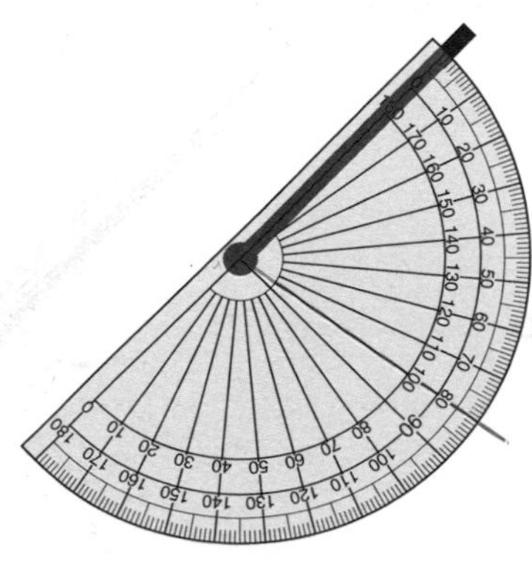

The angle is 80°.

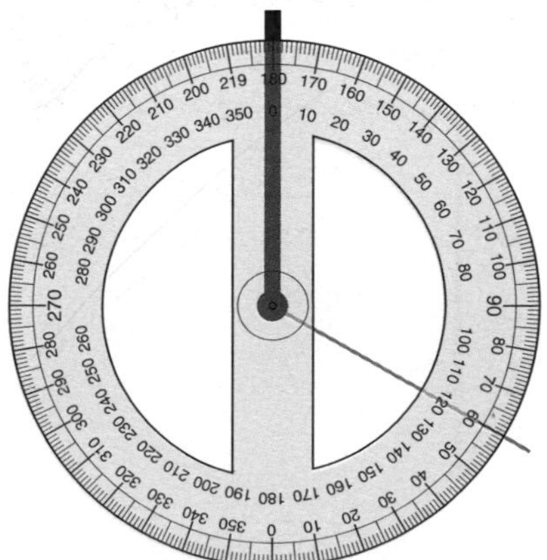

The angle is 120°.

Exercise 10:4

H 4 Measure these angles.
Count round in tens from 0°.
Write down your answers in degrees.

1

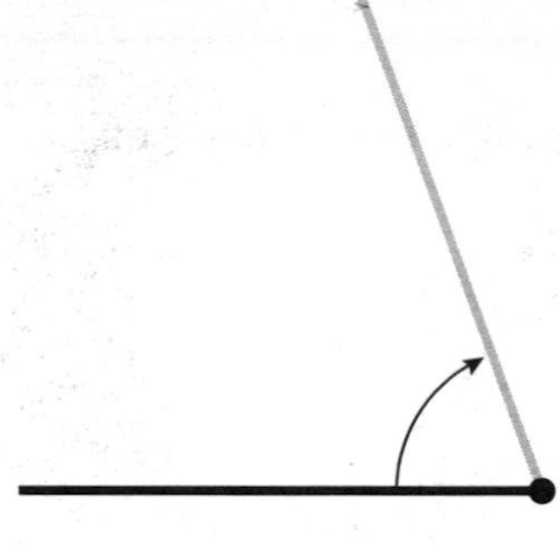

4

2

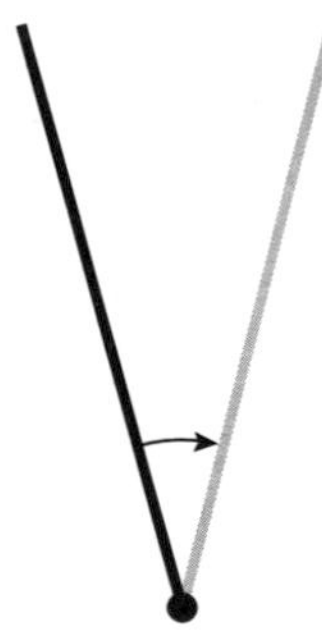

5

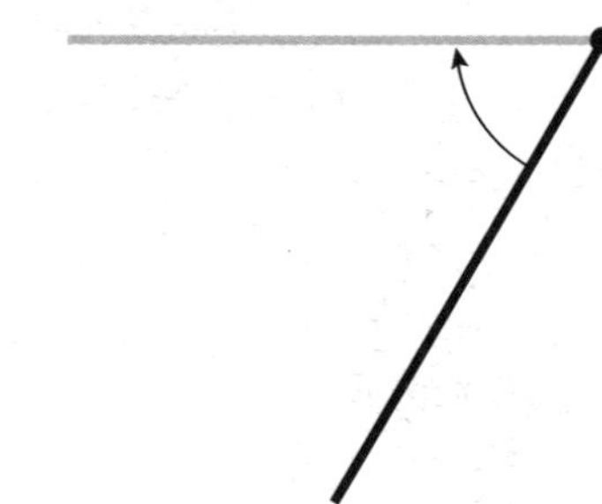

3

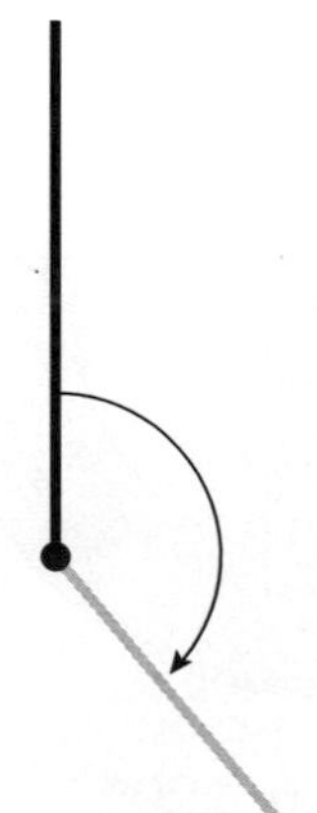

6

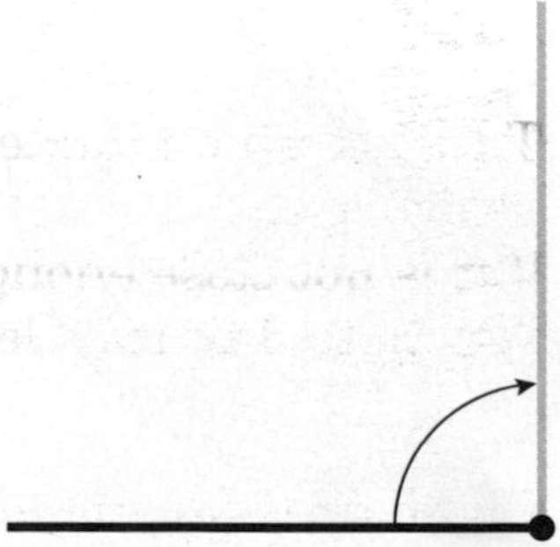

H 5, 6, 7, 8

Example Carolyn asks Sanjeev to draw an angle of 80°.
Sanjeev draws an angle which he estimates is 80° in his book.
He tries to remember what 90° looks like as he does it.

Carolyn checks it.
It is not close enough to 80°.
Sanjeev tries again.
They keep going until Sanjeev has drawn an angle which is very close to 80°.

Exercise 10:5

1 Try to draw an angle of 50°.
Measure it.
If it is not close enough, try again.
Try up to 3 times. Get as close as you can.

2 Try to draw an angle of 70°.
Measure it.
If it is not close enough, try again.
Try up to 3 times. Get as close as you can.

3 Try to draw an angle of 120°.
Measure it.
If it is not close enough, try again.
Try up to 3 times. Get as close as you can.

1, 2

3 Calculating with angles

You do not always find angles by measuring. You can calculate angles.

Exercise 10:6

W 1, 2 **1** **a** Carefully cut out the angles from your worksheets.

b Find the pairs of angles that fit together against a ruler.

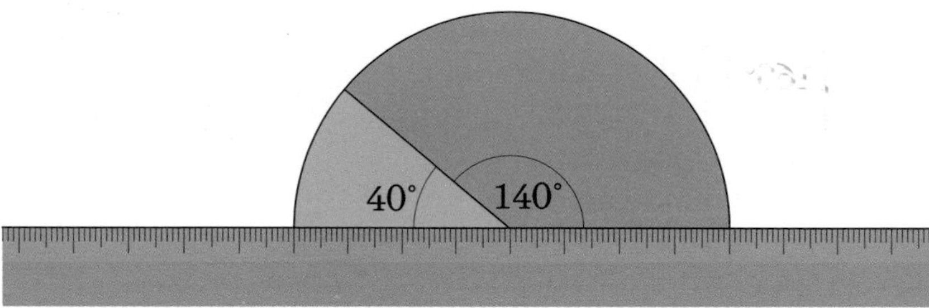

c Stick the pairs in your book.

d For each pair write down the sum:

………… + ………… = …………

The angles on a straight line always add up to 180°.

Example

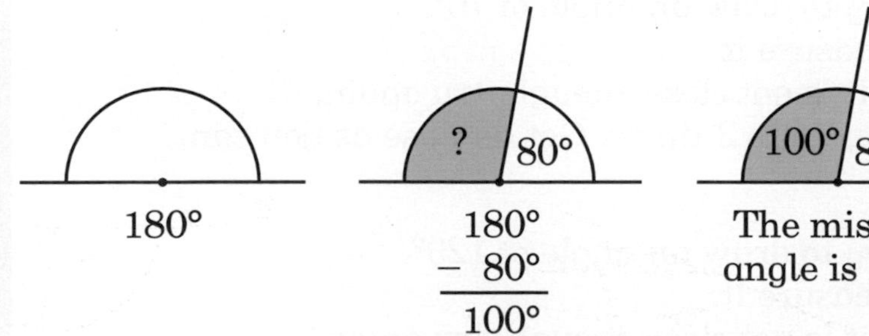

---

2 Find the missing angle.
Write the sum for each question.

a

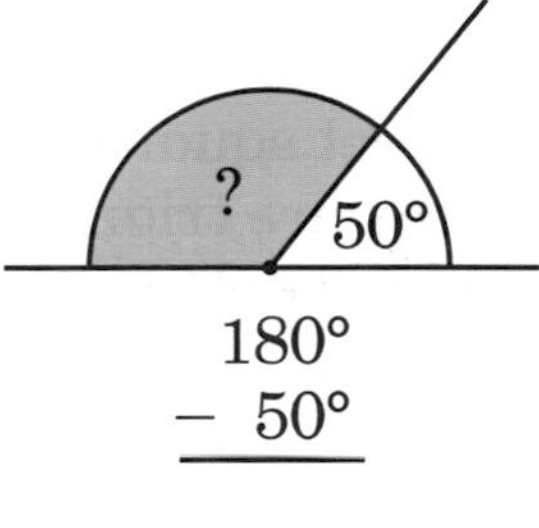

180°
− 50°

e

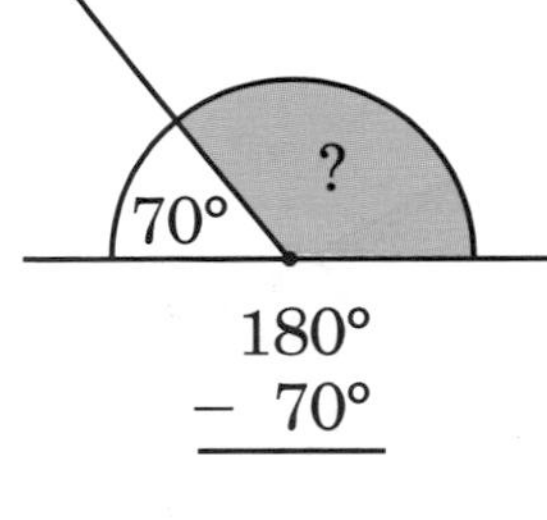

180°
− 70°

b

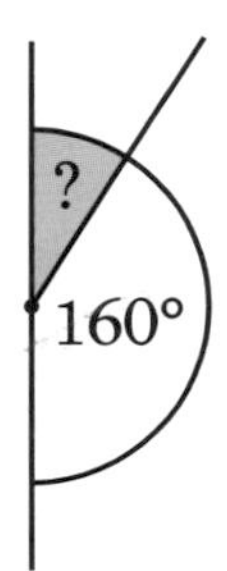

f

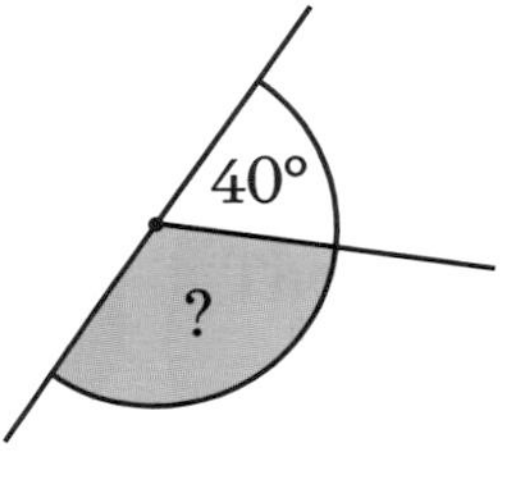

c

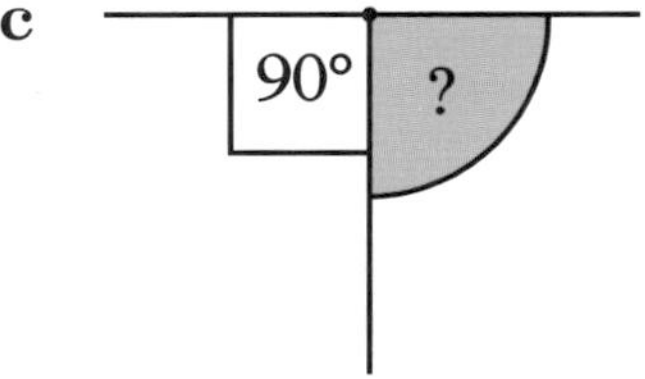

g

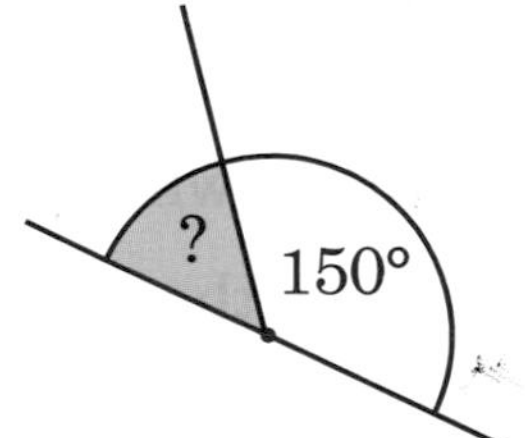

d

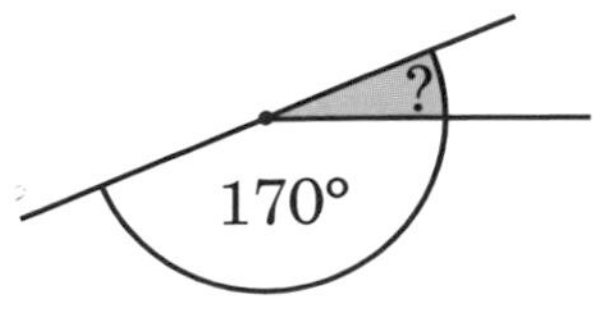

h

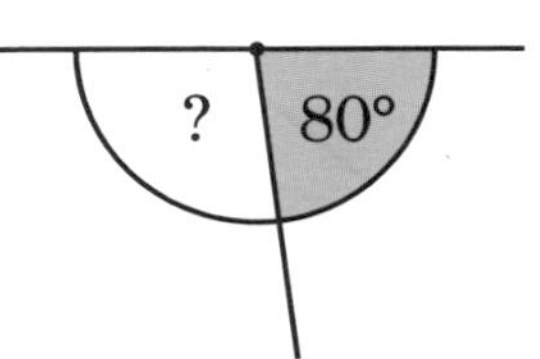

H 10

1 Use *your* right angle to check if *these* angles are right angles.
Write down either:

This is a right angle.
or This angle is bigger than a right angle.
or This angle is smaller than a right angle.

a

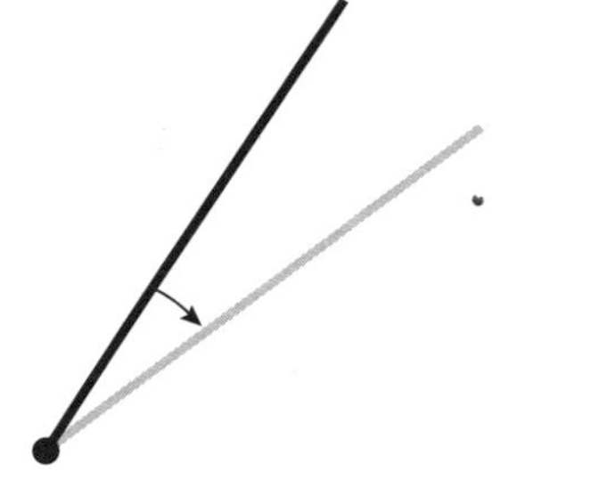

c

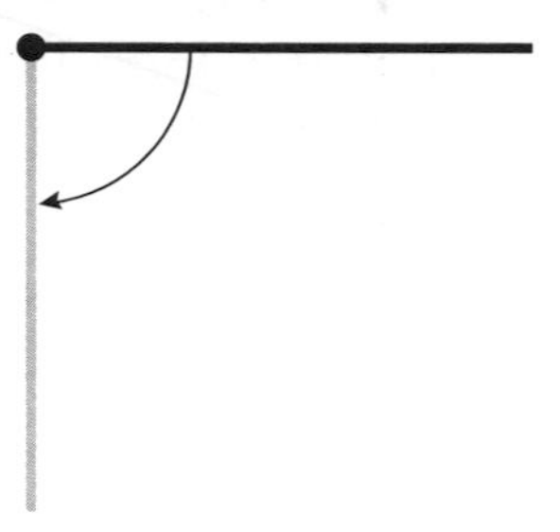

b

d

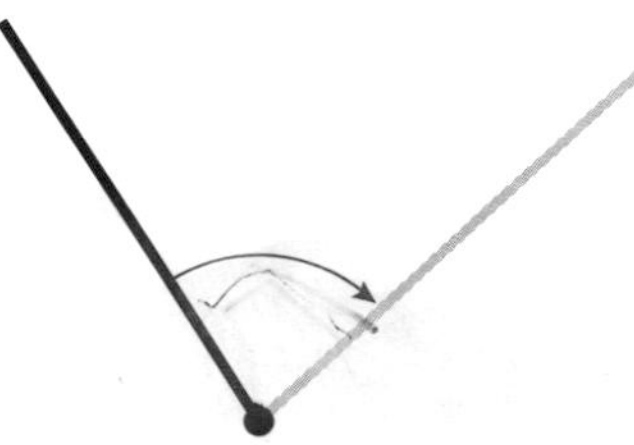

H 11

2 Use *your* right angle to check if *these* angles are acute angles or obtuse angles or right angles.
Write down either:

This angle is an acute angle.
or This angle is an obtuse angle.
or This angle is a right angle.

a

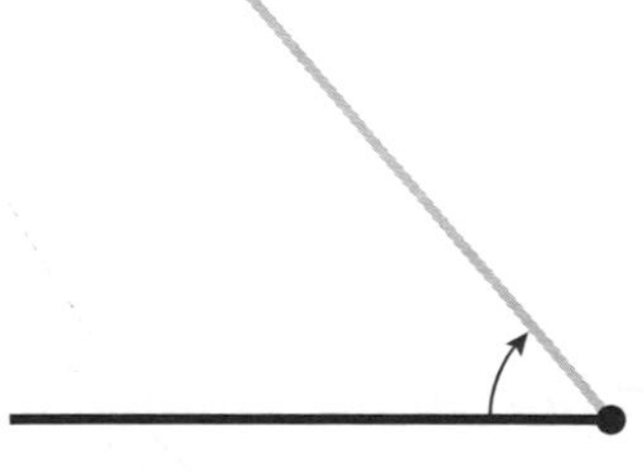

c

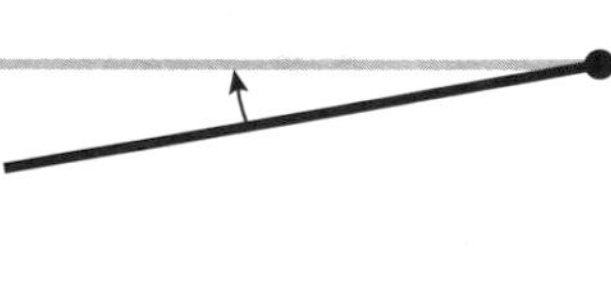

b

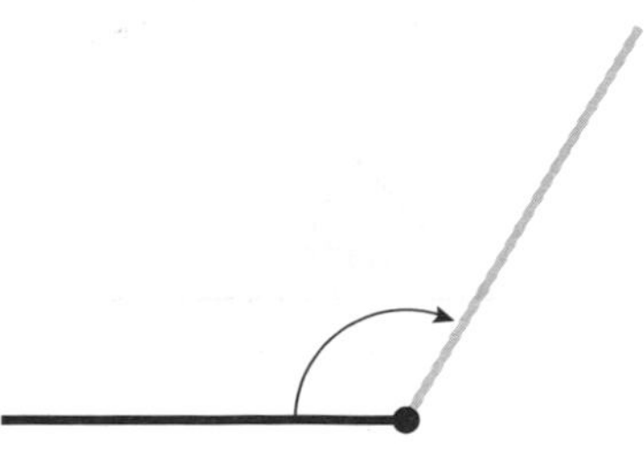

d

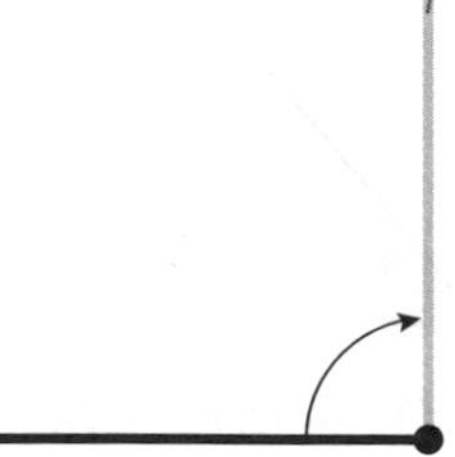

H 12 **3** Measure these angles.
Write down your answers in degrees.

a

c

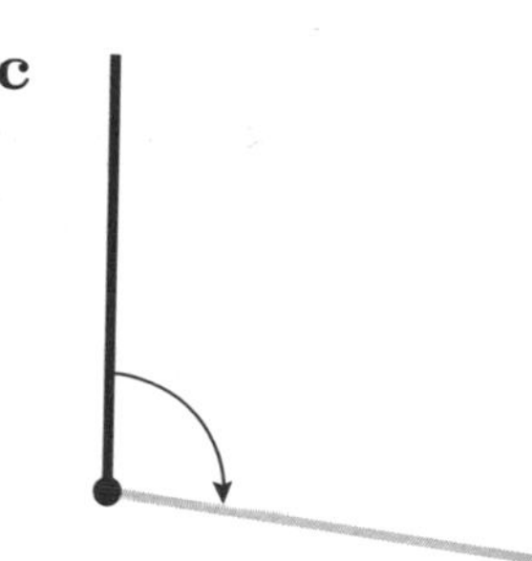

b

d

4 Try to draw an angle of 110°.
Measure it.
If it is not close enough, try again.
Try up to 3 times. Get as close as you can.

5 Find the missing angle.
Write down the sum for each question.

a

b

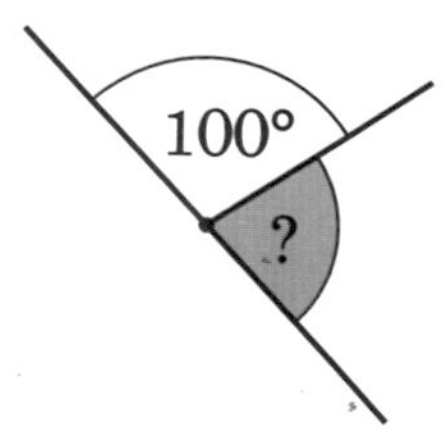

6 Find the missing angle.
Write down the sum for each question.

a

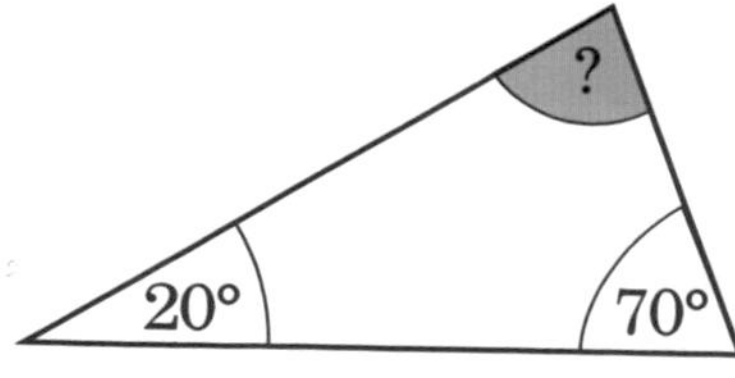

b

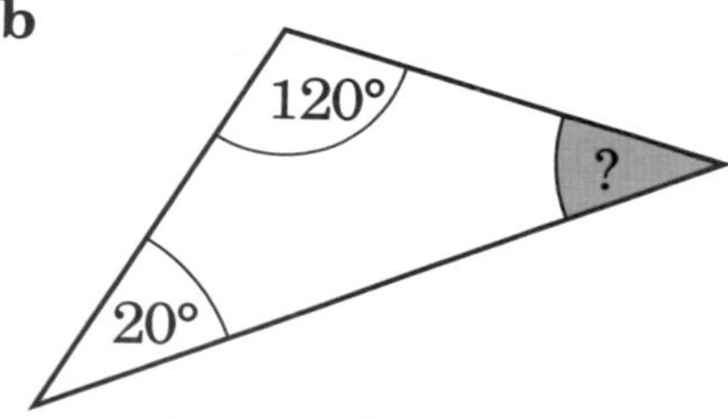

1 Introducing negative numbers

Bill the ice-cream man has to keep his ice-cream cold.

Bill's ice-cream is at −10 °C.

Example Negative numbers are less than nought: −2, −5, −32

Nought is not positive or negative 0

Positive numbers 7, +3, 25

(You do not always use +, so 3 is the same as +3)

Exercise 11:1

1 From this list: −7, +3, −4, 0, 5

a Write down the negative numbers.

b Write down the positive numbers.

2

1 degree above zero is 1°C ⟶ 1

Freezing point of water is 0°C ⟶ 0

1 degree below zero is −1°C ⟶ −1

−2

−3

−4

−5

°C

Copy this list.

Fill in the missing numbers.

2 degrees below zero is −2 °C
3 degrees below zero is ...
4 degrees below zero is ...
5 degrees below zero is ...

1, 2, 3

W1, 2 **3** Write down the temperatures shown by these thermometers.

a

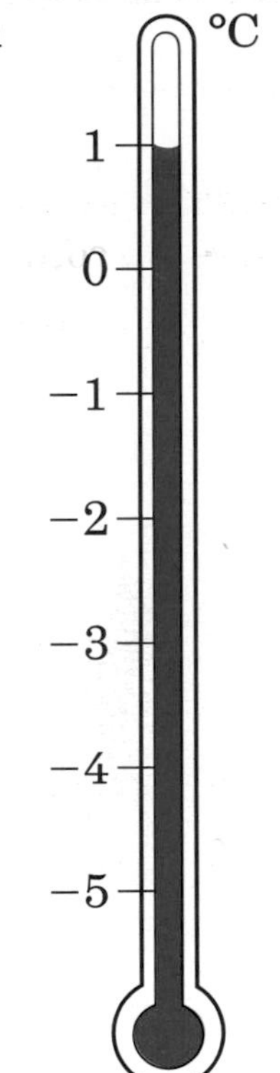

b

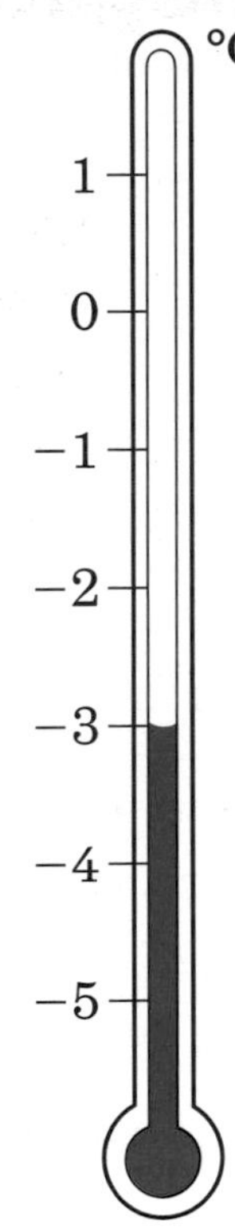

c

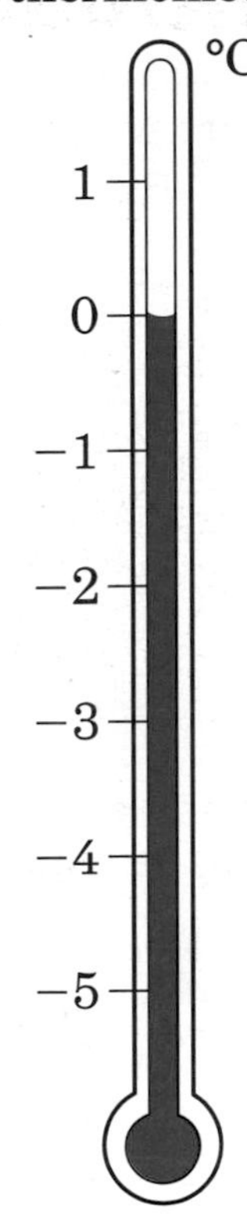

4 **a** Which is warmer, 4 °C or −3 °C?

b Which is warmer, 2 °C or −5 °C?

c Which is warmer, −4 °C or −2 °C?

d Which is warmer, 2 °C or −3 °C?

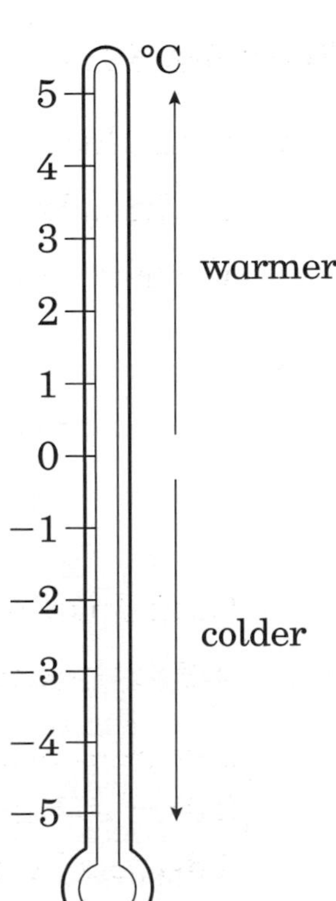

5 **a** Which is colder, 5 °C or −4 °C?

b Which is colder, 4 °C or −2 °C?

c Which is colder, −4 °C or −3 °C?

d Which is colder, −2 °C or −5 °C?

6 Look at the thermometer scale.

Write the numbers on the scale in order.
Start with the warmest.

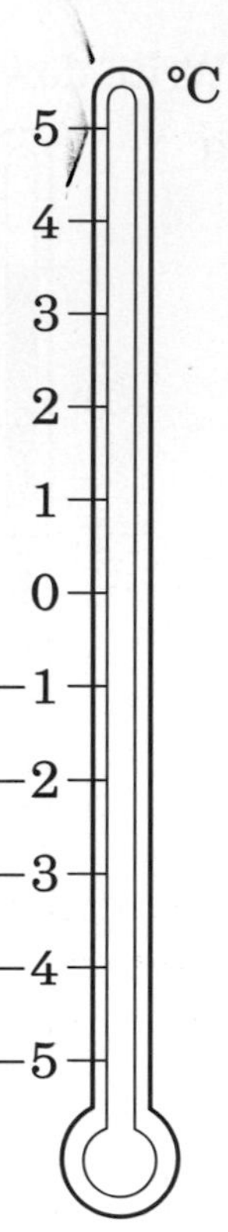

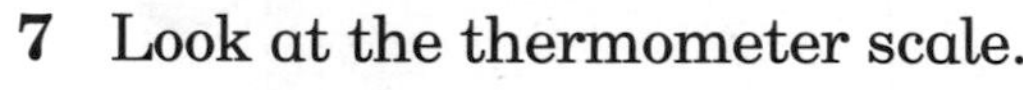

7 Look at the thermometer scale.

Write the numbers on the scale in order.
Start with the coldest.

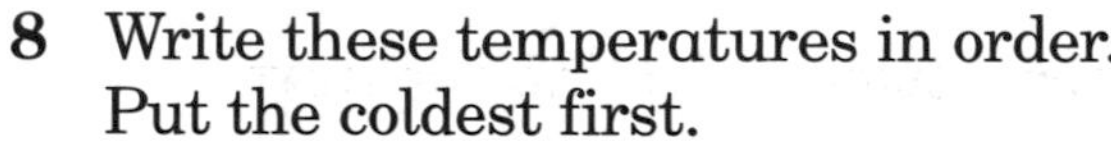

8 Write these temperatures in order.
Put the coldest first.

a −3 °C, 2 °C, 0 °C

b 4 °C, −1 °C, −3 °C

9 Write these temperatures in order.
Put the coldest first.

a −2 °C, −5 °C, 3 °C

b 0 °C, −1 °C, −4 °C

c 2 °C, 1 °C, −5 °C

10 Write these temperatures in order.
Put the coldest first.

a 5 °C, −6 °C, 0 °C

b 6 °C, 2 °C, −4 °C

11 Write the temperatures of these cities in order.
Write the coldest city first.

Edinburgh	−1 °C
Inverness	−4 °C
Glasgow	−2 °C
Manchester	2 °C

Example Write down the rule for this pattern.

Find the next two terms.

5, 3, 1, −1, −3

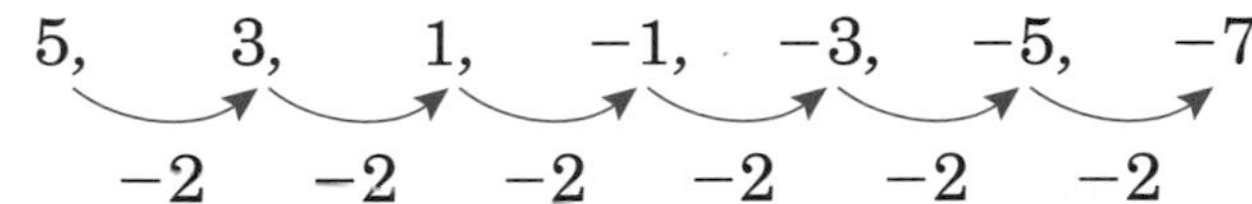

The rule is −2.

The next two terms are −5 and −7.

12 Write down the rule for each pattern and find the next two terms.
You could use a number line to help you.

a 4, 2, 0, −2, −4, ..., ...

b 2, 1, 0, −1, −2, ..., ...

c 9, 6, 3, 0, −3, ..., ...

d 1, −1, −3, −5, −7, ..., ...

e 10, 8, 6, 4, 2, ..., ...

Example One night the temperature is −2 °C.

The next day the temperature is 3 °C.

What is the difference between the day and night temperatures?

The difference is 5°C.

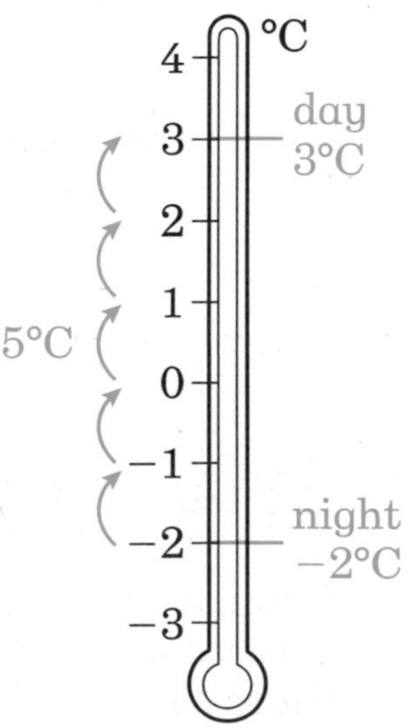

W3 **13** What is the difference between these day and night temperatures?

a

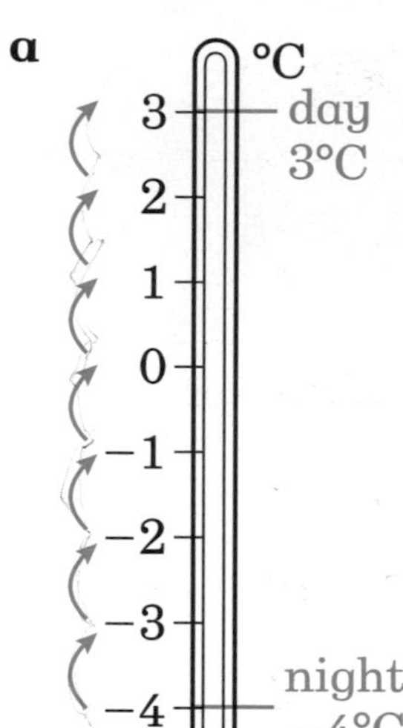

c

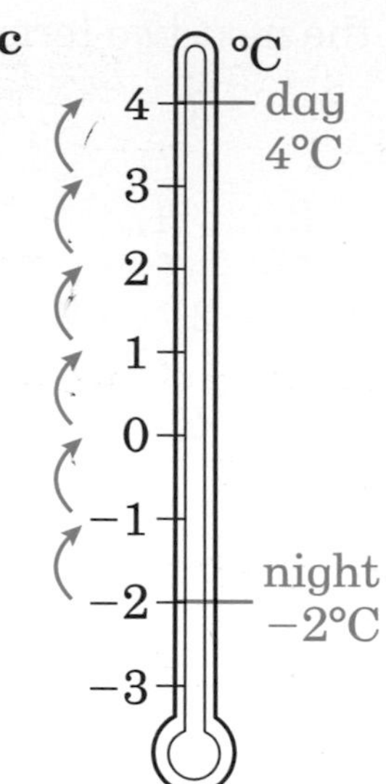

e

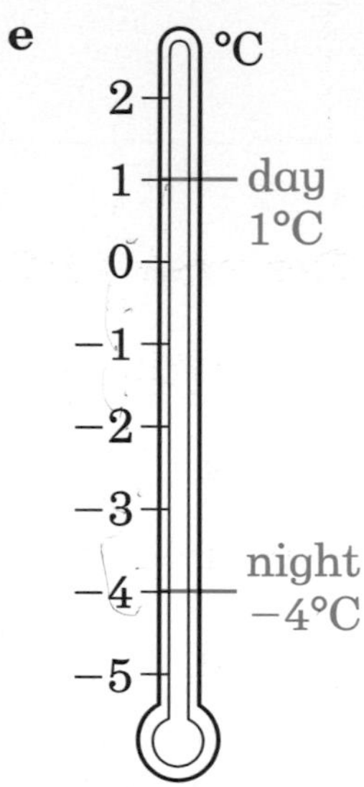

b

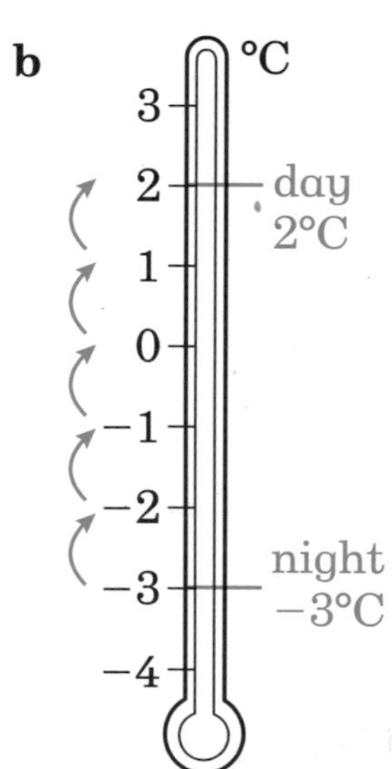

d

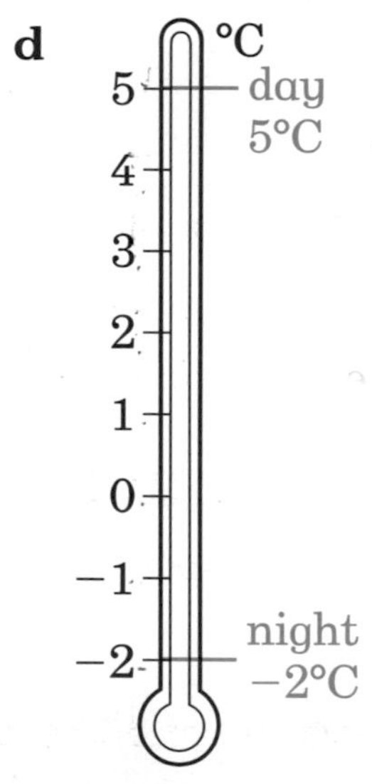

f

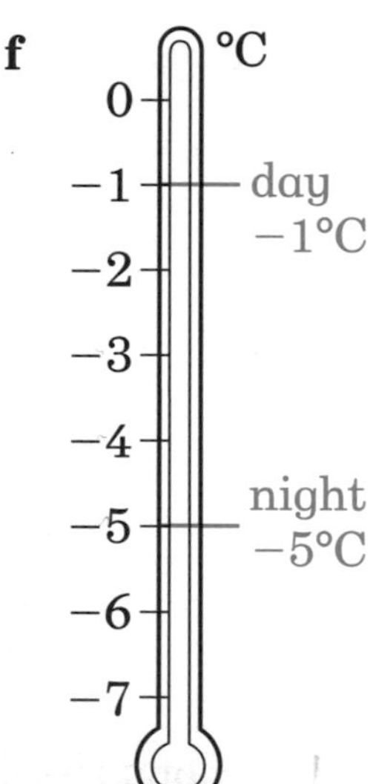

14 One day the temperature is 3 °C.

That night the temperature is −5 °C.

What is the difference in temperature?
You could use a number line to help you.

15 One day the temperature is 4 °C.

That night the temperature is −3 °C.

What is the difference in temperature?

Example Work this out using a calculator: 5 − 9

Keys to press: 5 −   −4

Exercise 11:2

Write down the answers shown on your calculator.

1 4 − 7 **4** 0 − 5 **7** 16 − 18

2 1 − 5 **5** 8 − 5 **8** 17 − 13

3 3 − 6 **6** 12 − 16 **9** 20 − 25

Example One day the temperature is 4 °C.

That night the temperature fell by 7 °C.

Find the night temperature.

Keys to press:   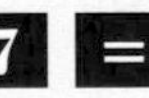 = −3 Answer −3 °C

W 4, 5 **10** Copy the table. Find the night temperatures.

	Day temperature	Fall in temperature	Night temperature
a	4 °C	6 °C	−2 °C
b	2 °C	4 °C	…
c	1 °C	4 °C	…
d	2 °C	7 °C	…
e	5 °C	5 °C	…
f	3 °C	6 °C	…

G 4

2 Using negative numbers

Jane and Peter are going in the lift.

They are on the third floor.

They are going to the first floor.

They will go down 2 floors.

Exercise 11:3

1 Here is the control panel of a lift.

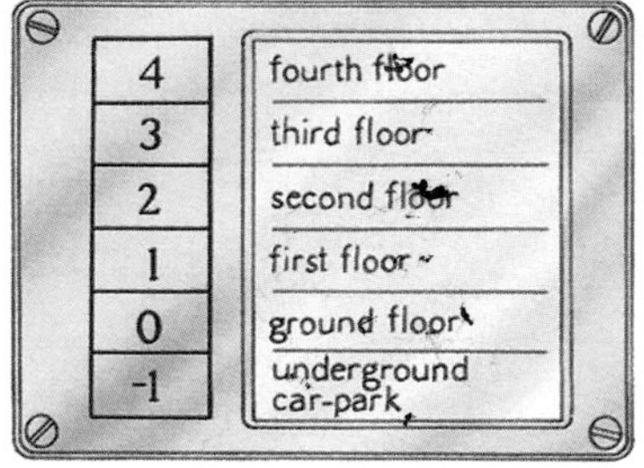

a What number is used for the ground floor?

b Where do you go if you press [−1]?

c You go from the second floor to the ground floor. How many floors do you go down?

d You go from the fourth floor to the first floor. How many floors do you go down?

e You go from the ground floor to the car park. How many floors do you go down?

f You go from the third floor to the car park. How many floors do you go down?

G 5 **g** You go from the second floor to the car park. How many floors do you go down?

2 How many metres is it from:
(the first one is done for you)

a the girl to the bird? +2 metres

b the girl to the crab?

c the girl to the jellyfish?

d the bird to the fish?

e the crab to the bird?

f the jellyfish to the bird?

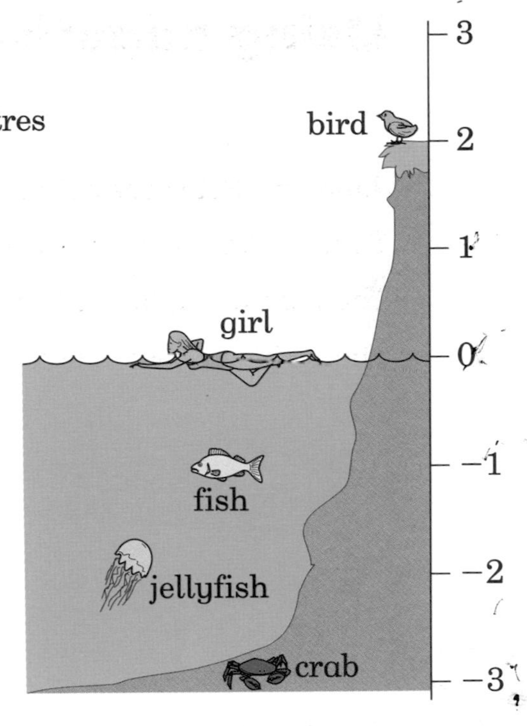

3 How many metres is it from:

a sea level to the helicopter?

b sea level to the diver?

c sea level to the wreck?

d the diver to the wreck?

e the helicopter to the wreck?

f the helicopter to the diver?

g the diver to the top of the cliff?

h the wreck to the top of the cliff?

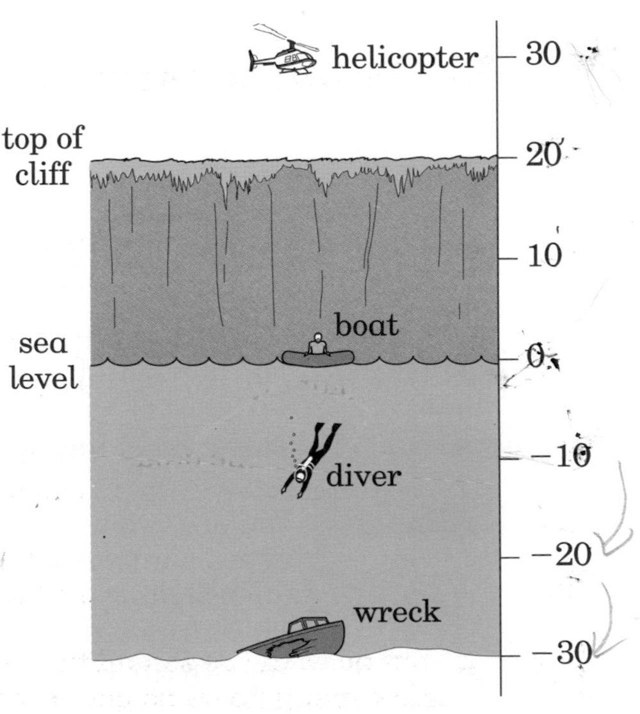

3 Co-ordinates

Tariq is using this map.

Tariq wants to find the campsite Ⱥ

It is at (4, 3)

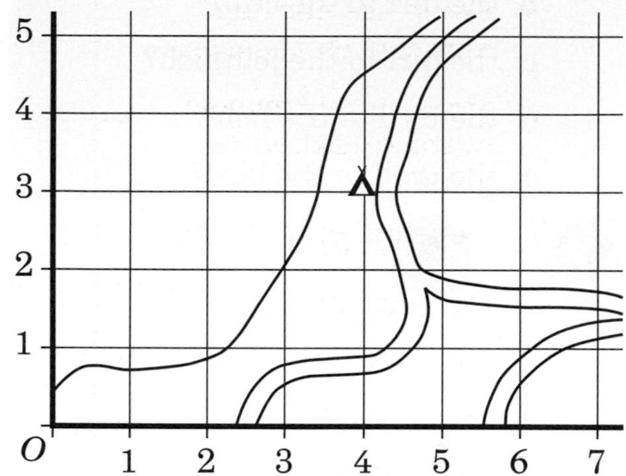

Axes You draw a horizontal line and a vertical line known as **axes**.

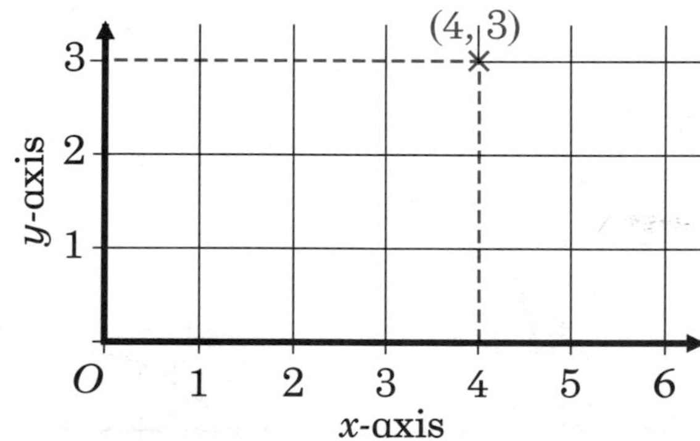

***x*-axis** The horizontal line is called the ***x*-axis**.

***y*-axis** The vertical line is called the ***y*-axis**.

Co-ordinates You use two numbers to mark a point.
These numbers are called **co-ordinates**.
Co-ordinates are written like this (4, 3)
This means it is across 4 and up 3.
You always write the across number first.

Exercise 11:4

6 **1** **a** Use squared paper. Copy these axes.

b Fill in the missing numbers on the axes.

c Draw these symbols at the co-ordinates given.

(3, 5)

(6, 2)

(2, 3)

(5, 4)

(1, 5)

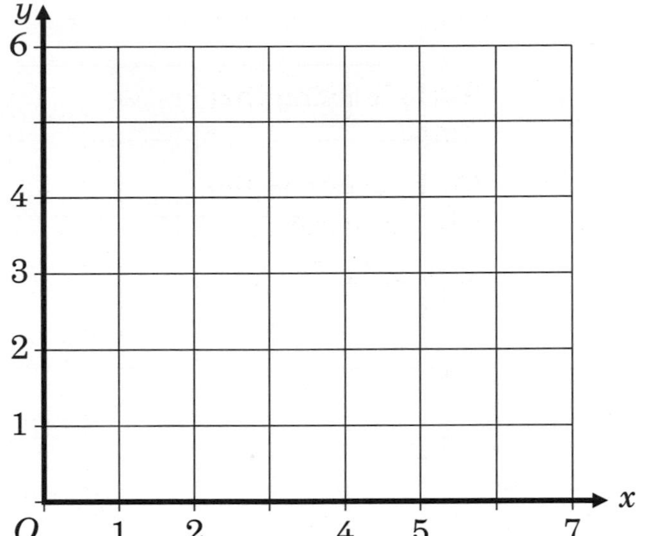

6 **2**

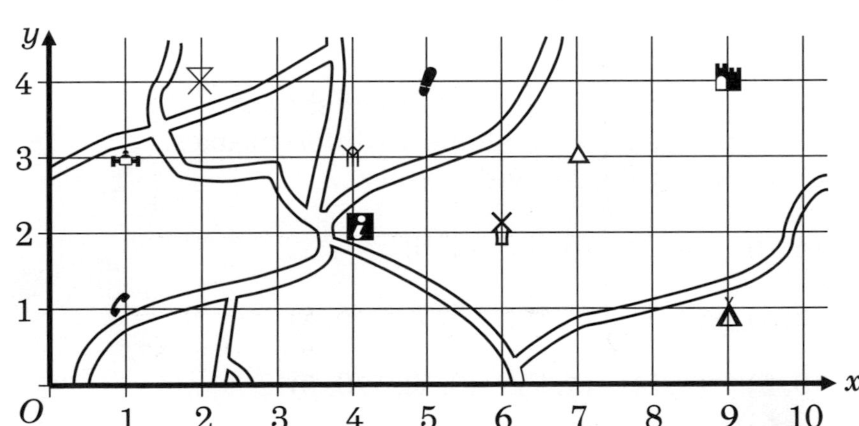

Write down the co-ordinates of these.
The first one is done for you.

a Country Park (4, 3)
b Nature trail
c Camp site
d Picnic site
e Youth hostel
f Castle
g Motor racing
h Information
i Telephone
j Windmill

W 7 **3** You need 4 grids like this:

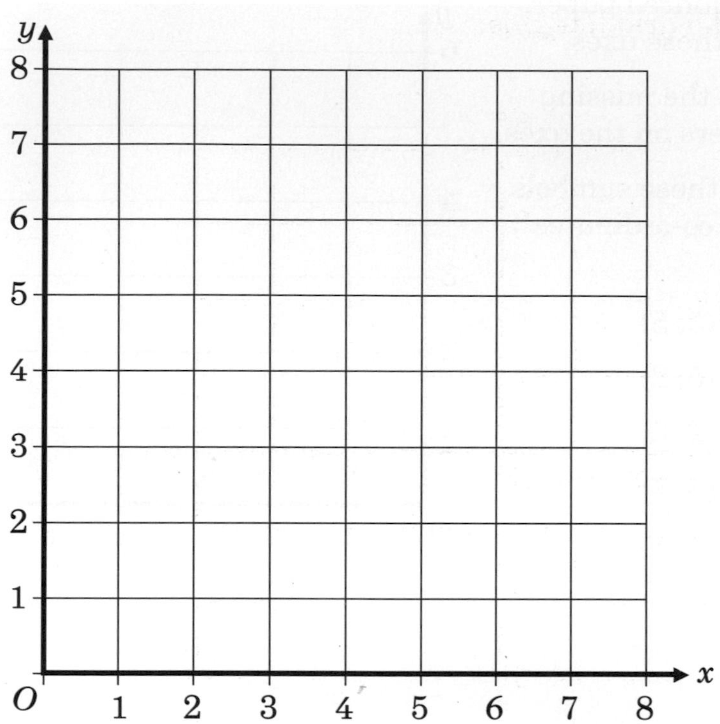

These co-ordinates draw 4 different shapes.
Plot the points for each shape on different axes.
Join them up in order with a ruler.

a (4, 0), (8, 0), (8, 3), (4, 0)

b (6, 4), (7, 7), (6, 8), (5, 7), (6, 4)

c (0, 5), (4, 5), (4, 7), (0, 7), (0, 5)

d (2, 0), (4, 2), (2, 4), (0, 2), (2, 0)

6 **e** Now label the shapes with their names.
Choose from: rectangle, kite, triangle, square.

 8

4 These two number lines (axes) have negative numbers.
The lines cross at (0, 0).

You need a grid like this:

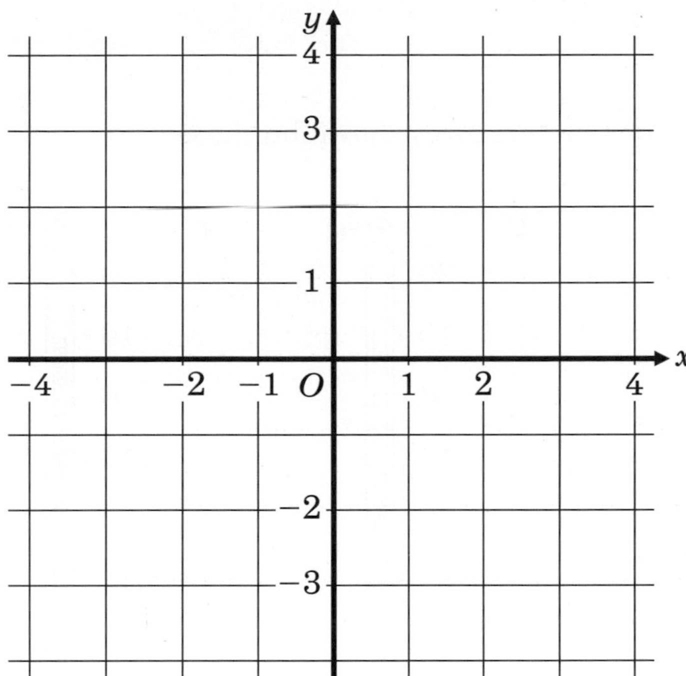

Fill in the missing numbers on the axes.

Plot these points.
Remember: across first, then up or down.

Join the points with a ruler as you go.

a (0, 0), (3, 3), (2, 0)

b (0, 0), (−3, 3), (−2, 0)

c (0, −1), (3, −4), (2, 0)

d (0, −1), (−3, −4), (−2, 0)

1 From this list: −1, 2, 0, +6, −5

a write down the positive numbers,

b write down the negative numbers.

2 Write down the temperatures shown by these thermometers.

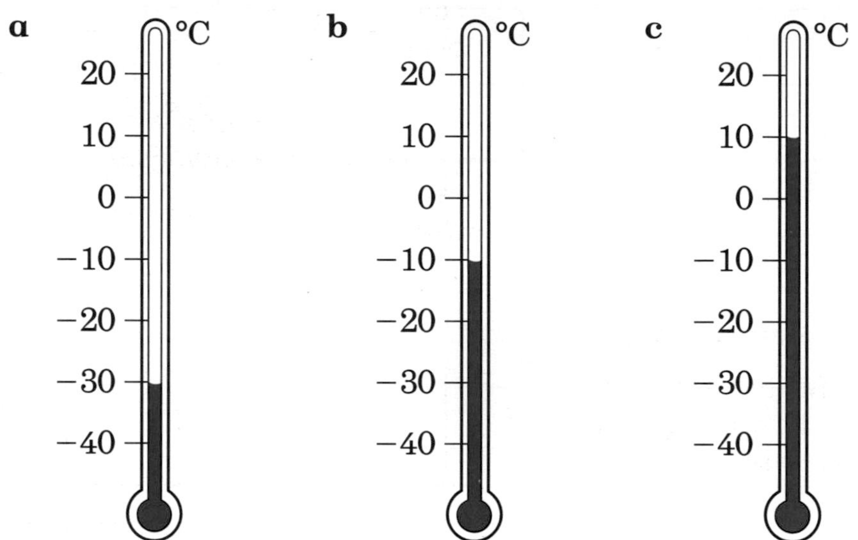

d Which is colder, −5 °C or −1 °C?

3 Write these temperatures in order.
Put the coldest first.

a 0 °C, −3 °C, 2 °C

b −1 °C, −5 °C, −3 °C

4 One day the temperature is −3 °C.

Later that day the temperature falls to −5 °C.

How many degrees has the temperature fallen?

5 How many metres is it from:

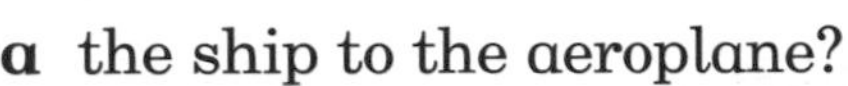

a the ship to the aeroplane?

b the ship to the bottom of the sea?

c the ship to the submarine?

d the aeroplane to the submarine?

e the bottom of the sea to the submarine?

f the bottom of the sea to the aeroplane?

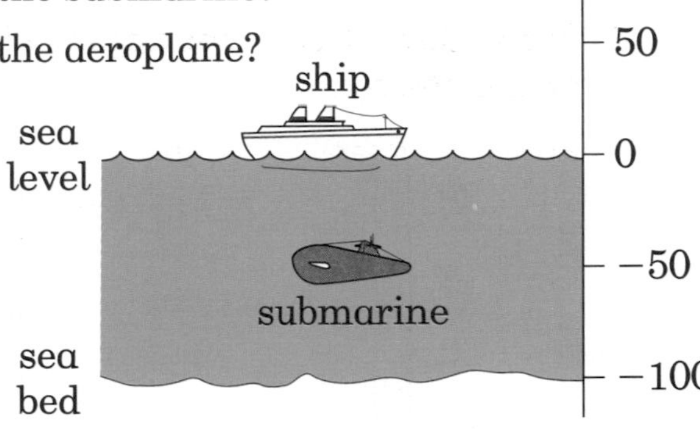

6 Use squared paper.

Draw an x-axis from 0 to 13 and a y-axis from 0 to 13.

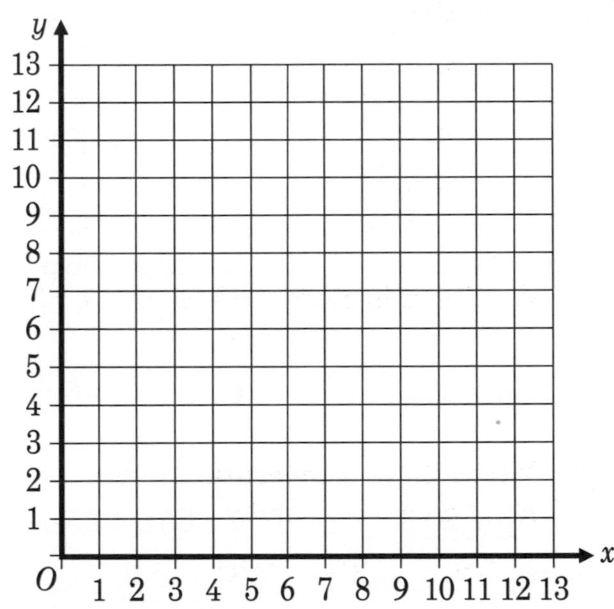

Plot these points.

Join them up in order with a ruler.

a (7, 8), (4, 11), (6, 11), (9, 8)

b (9, 8), (13, 8), (13, 7), (9, 7), (6, 4)

c (7, 8), (0, 8), (0, 7), (7, 7), (4, 4)

d (6, 4), (13, 4), (13, 3), (6, 3), (9, 0)

e (4, 4), (0, 4), (0, 3), (4, 3), (7, 0)

Money 2

Exercise 1

Example
20 p + 40 p = 60 p

Find the total in each of these questions.

1 10 p + 20 p

2 20 p + 20 p

3 50 p + 20 p

4 50 p + 40 p

5 60 p + 10 p

6 70 p + 20 p

7 80 p + 20 p

8 60 p + 50 p

9 70 p + 40 p

10 80 p + 40 p

Exercise 2

Example
25 p + 15 p = 40 p

Find the total in each of these questions.

1 5 p + 15 p

2 5 p + 35 p

3 15 p + 15 p

4 25 p + 35 p

5 25 p + 55 p

6 35 p + 45 p

7 35 p + 55 p

8 45 p + 55 p

9 45 p + 75 p

10 75 p + 85 p

Exercise 3

Example
12 p + 29 p = 41 p

Find the total in each of these questions.

1 11 p + 19 p

2 13 p + 18 p

3 17 p + 26 p

4 24 p + 33 p

5 36 p + 56 p

6 42 p + 39 p

7 53 p + 61 p

8 68 p + 72 p

9 77 p + 77 p

10 98 p + 28 p

12 Units of length and scale drawing

CORE

1 **An introduction to units of length**

2 **Scale drawings**

3 **Finding lengths**

QUESTIONS

HELP YOURSELF **Time 2**

The Romans measured in miles. 1 mile was 1000 paces. They counted two steps – left, right – as one pace.

But hands, feet and paces vary. Edward I was the first person in England to set out a system. In 1305, he had standard measures for people to compare, but the units were still complicated. Different measures were in use in different countries. Something better was needed.

In 1795, the French Academy of Science decided to solve the problem. A survey was made of the length of a line from the North Pole, through France to the Equator. This distance was divided into 10 million parts. Each part was called a metre. This was the start of the metric system.

1 An introduction to units of length

In ancient times people used parts of the body to measure length.

thumb	(middle joint to tip)
hand	(sideways across wide part)
span	(hand stretched wide, thumb up to little finger tip)
foot	(length to tip of big toe)
pace	(length of one step)

You can use these units to measure lots of different things.

Exercise 12:1

1 You want to measure the width of this book.
Which of these units would you use?

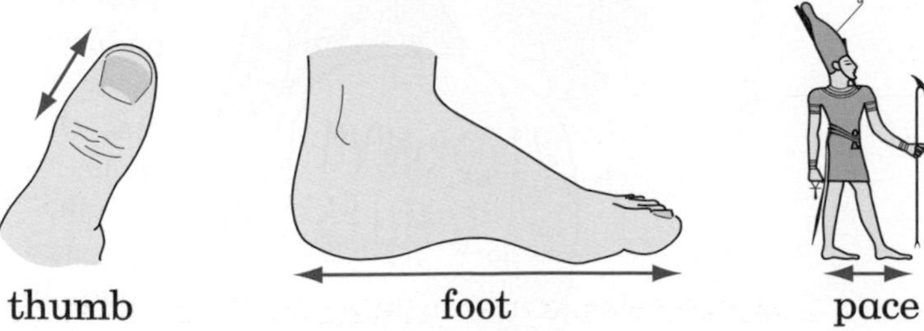

2 Which of these units would you use to measure:

a the length of this book?

b the width of your exercise book?

c the length of the classroom?

d the length of your playground?

e the width of the classroom door?

f the width of the classroom?

g the width of your desk?

h the length of your pen?

i the width of the blackboard?

Joshua decided to measure this book in thumbs.
He said the length is 7 thumbs and the width is 5 thumbs.

Sally measured the book in thumbs.
She said the length is 10 thumbs and the width is 8 thumbs.

Measure the width and length of this book in thumbs.
Do you agree with Joshua or Sally?

Joshua and Sally were both right, so were you!
This is because people have different size thumbs.

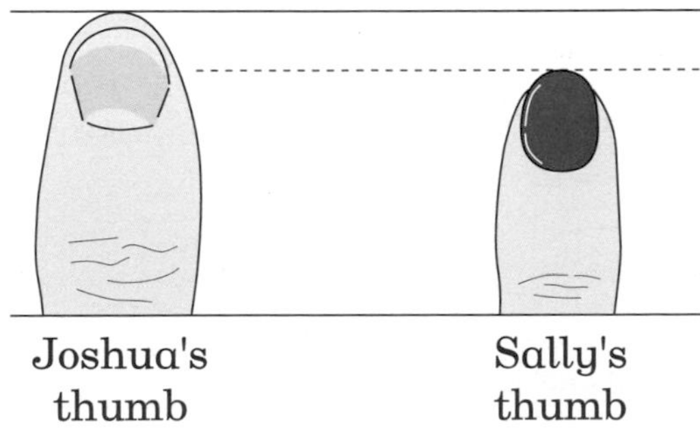

We all need to use the same units of length.

Millimetres **Millimetres** are used for measuring very small things.

This bug is 1 mm long. 

Centimetres **Centimetres** are used for measuring slightly bigger things.

This beetle is 1 cm long. 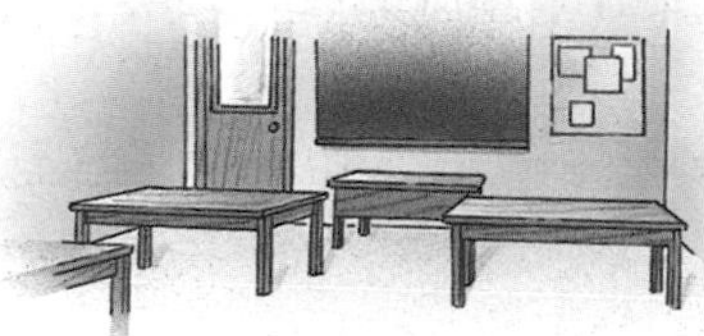

Metres **Metres** are used for measuring big things.

You would measure the length of your classroom in metres.

Kilometres **Kilometres** are used for measuring long distances.

You would measure the distance from London to your school in kilometres.

Exercise 12:2

1 Which of these units, millimetres (mm), centimetres (cm), metres (m) or kilometres (km), would you use to measure:

a the width of this book?

b the length of this book?

c the width of your exercise book?

d the length of the classroom?

e the length of the playground?

f the width of the classroom door?

g the width of the classroom?

h the width of your desk?

i the length of your pen?

j the width of the blackboard?

k the distance from your house to school?

l the length of a football pitch?

m the distance from your school to Manchester?

n the length of your little finger?

o the length of your foot?

Exercise 12:3

1 This ruler is marked in centimetres.

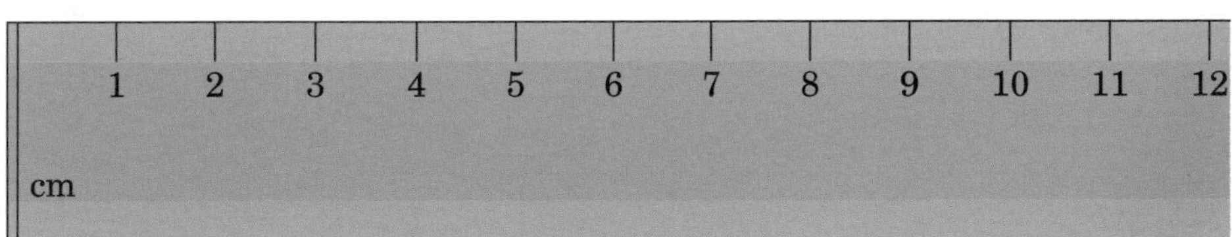

Use the ruler to help you *estimate* the lengths of these insects.
Remember that an estimate is a sensible guess.

a

b

c

d

e

f

Exercise 12:4

1 This line is 5 cm long:
Estimate the lengths of these lines in centimetres.

a

b

c

d

e

f

g

h

To measure accurately you need to use a ruler.

George is measuring the lines in Exercise 12:4.
Look carefully at how he uses his ruler.
He makes sure that the end of the line is at 0 on his ruler.

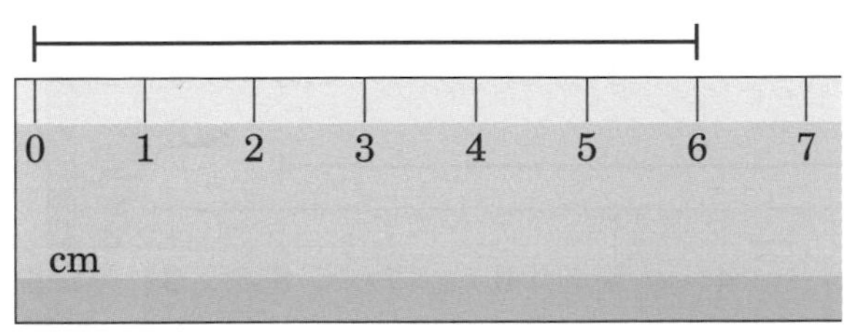

This line is 6 cm.

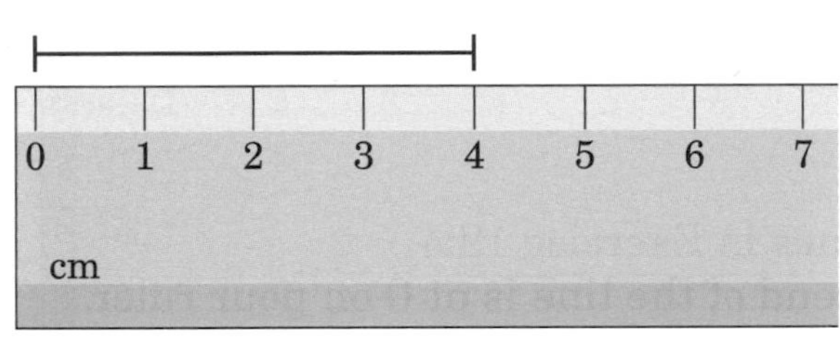

This line is 4 cm.

Exercise 12:5

1 Write down the lengths of these lines.

a

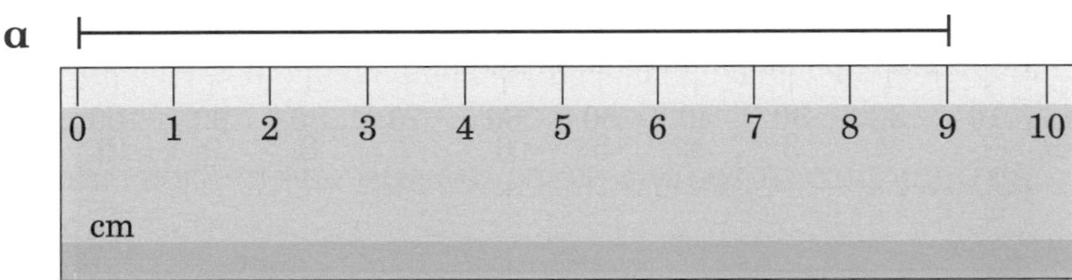

b

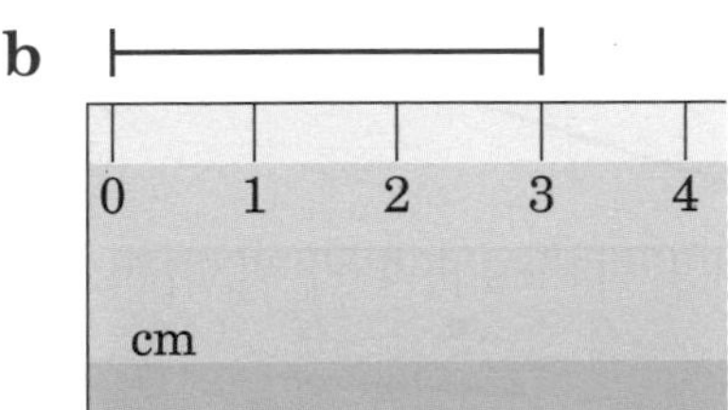

c

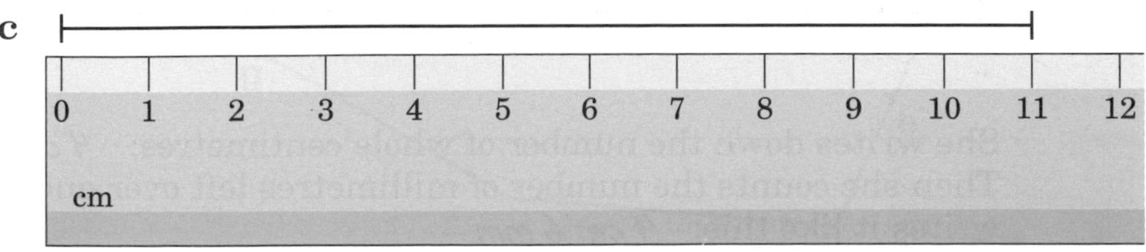

This is Val's ruler.

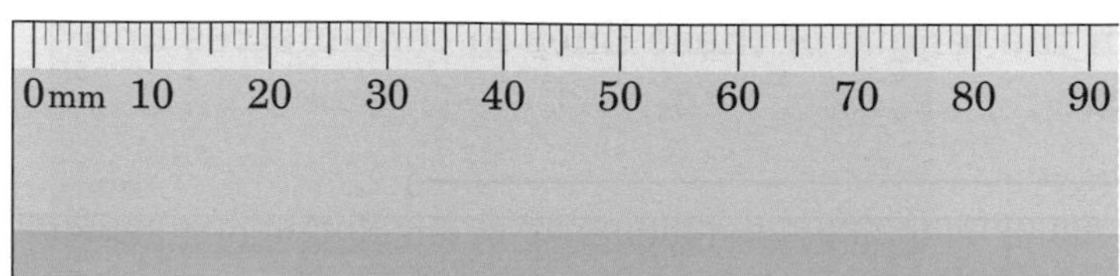

Val measures this line using her ruler.

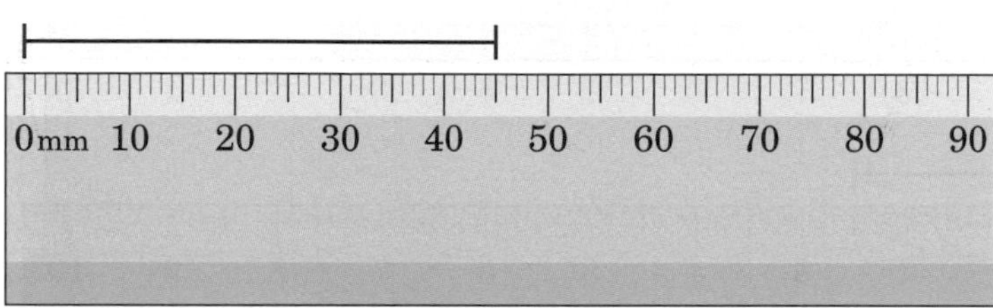

She writes down the number of millimetres: *45 mm*

Exercise 12:7

1 Write down the lengths of these lines in millimetres.

a

b

c

d

e

---

10 millimetres are equal to 1 centimetre.

Example Julie and Val measure this line.

Julie says it is *5 cm 4 mm*
Val says it is 54 mm.

Exercise 12:8

1 Measure these lines and write down their lengths.
Write them in centimetres and millimetres like Julie
and then in millimetres like Val.

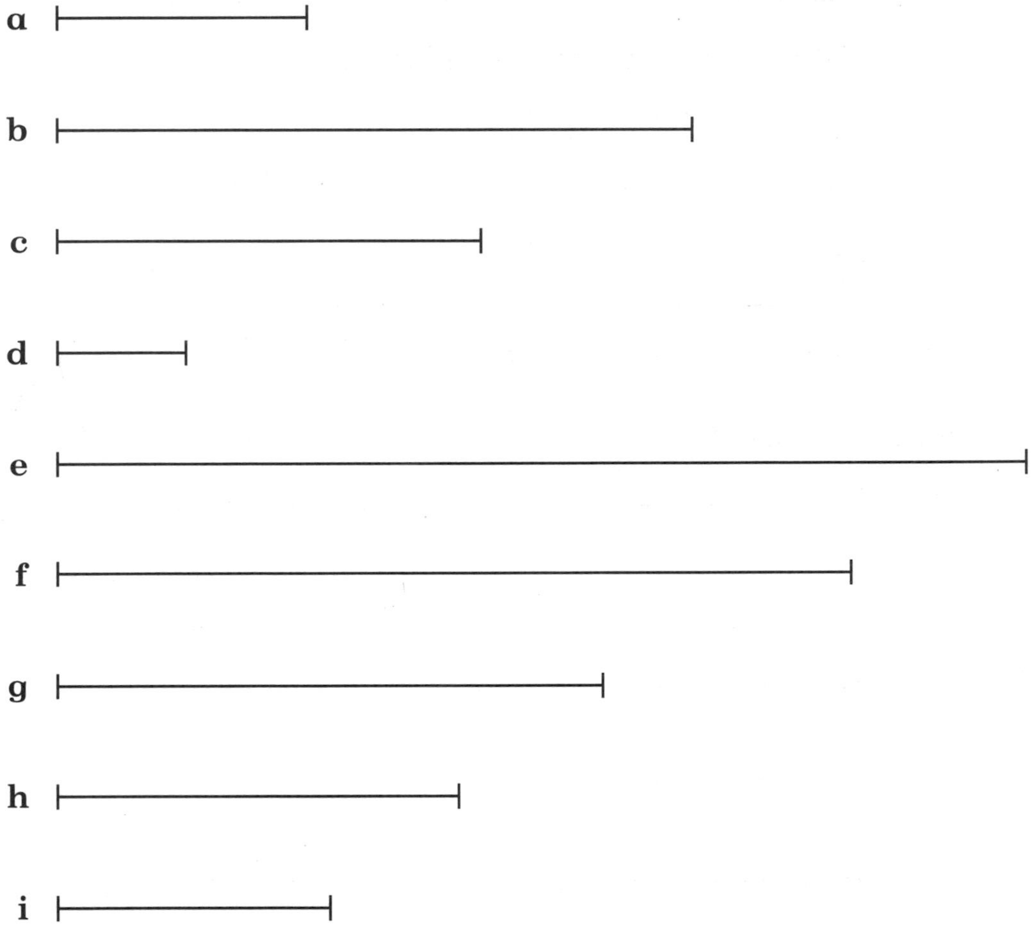

a

b

c

d

e

f

g

h

i

Example Julie is doing her homework.
Her teacher has asked her to measure the length
and width of some stamps in millimetres.
She measures each stamp with her ruler, then
changes her measurements to millimetres.

Julie measures this stamp with her ruler:

0 cm 1 2 3 4 5 6 7

3 cm 2 mm = 32 mm

Exercise 12:9

1 Measure each of these stamps in millimetres.

a

b

c

d

e

f

2 Scale drawings

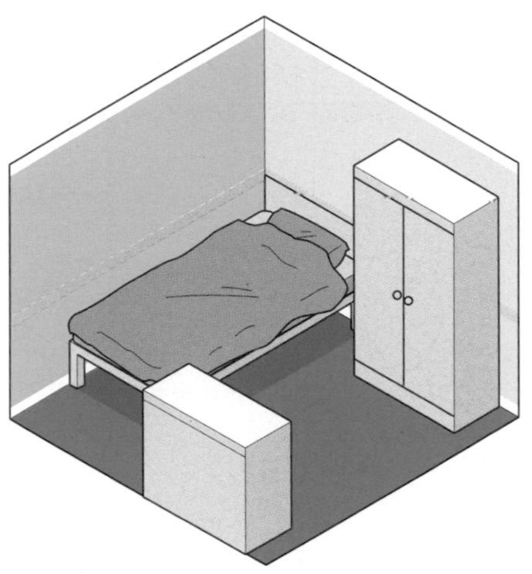

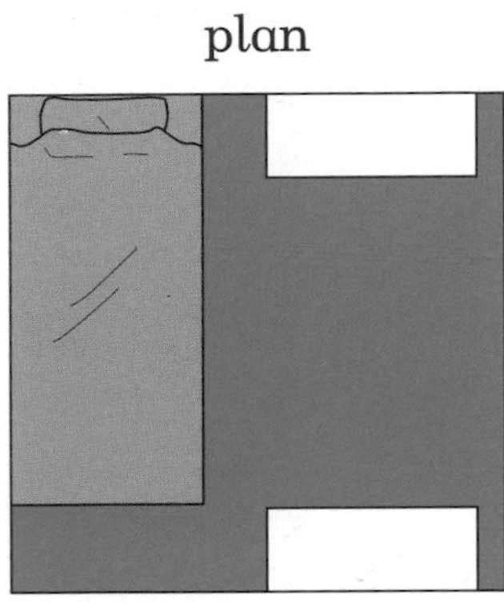

Scale drawing	A **scale drawing** is an accurate plan of something.
Scale	The **scale** helps you calculate real distances from the plan.

Exercise 12:10

1 You need a worksheet for this exercise.

Peter has bought some new furniture for his bathroom.

He has a new bath, sink and toilet.
He also wants to keep his laundry basket in the bathroom.

bath

toilet

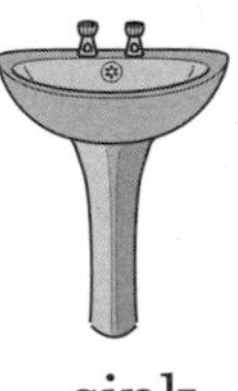

sink

laundry basket

Peter has drawn all the furniture to scale and he has a plan of his bathroom.

Put the furniture into Peter's bathroom in the best place you can.

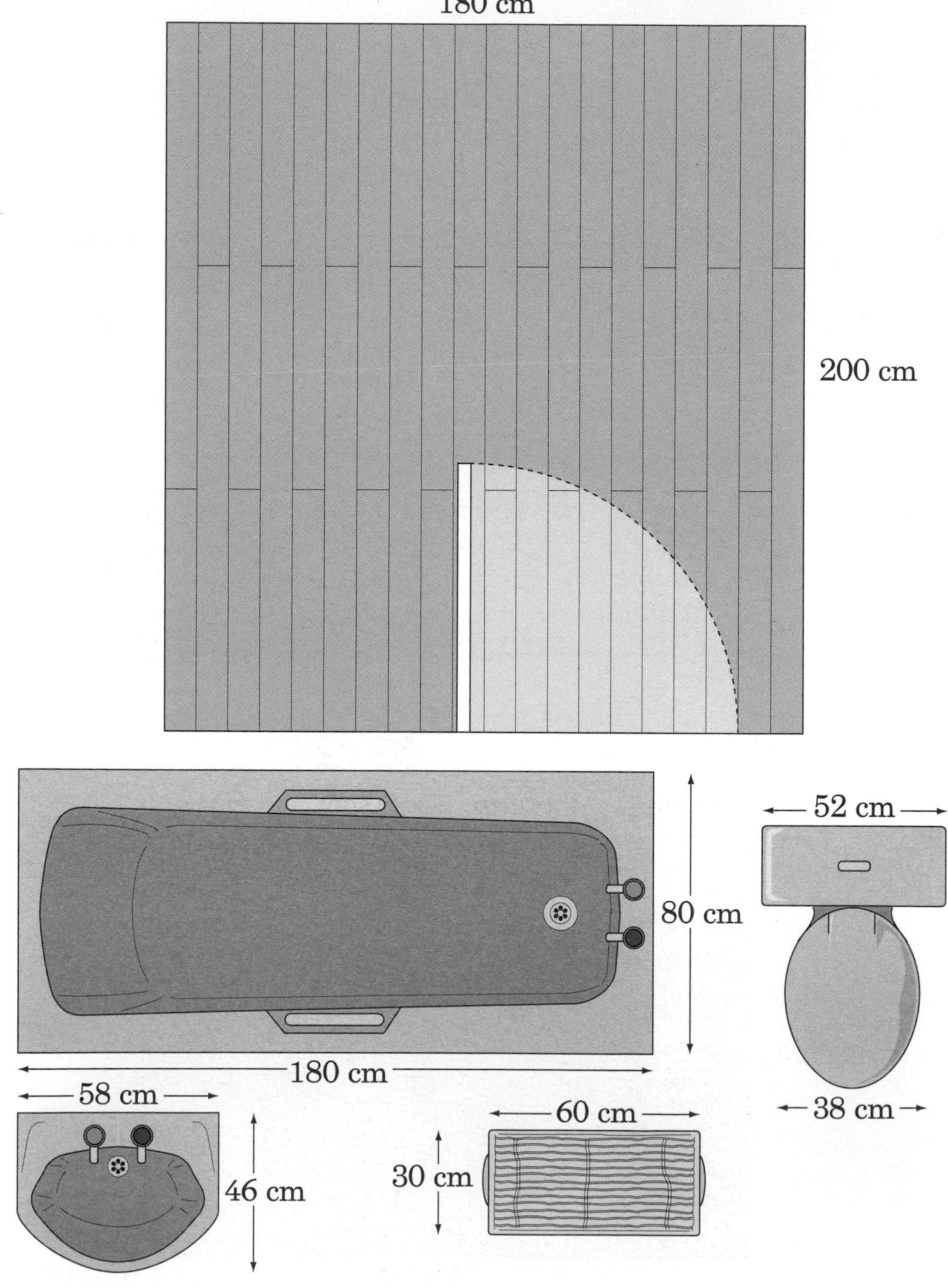

Exercise 12:11

W 2 You need a worksheet for this exercise.
Molly has bought some new units to put in her kitchen.

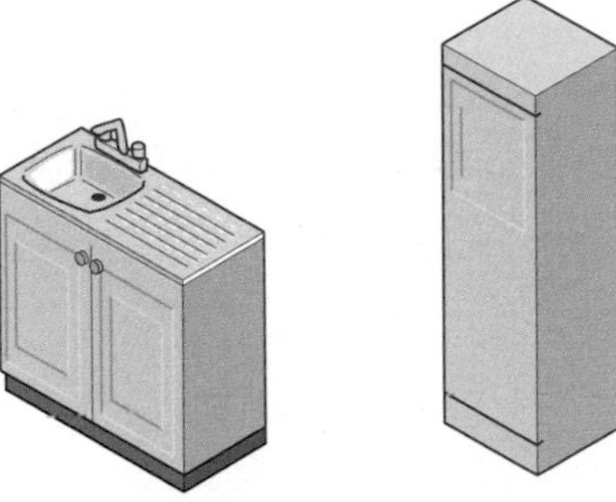

She wants to put them into her kitchen. She already has a cooker, washing machine and a refrigerator.

Molly has drawn all the units to scale and she has a plan of her kitchen.

Put the units into Molly's kitchen in the best place you can. Take care not to put the tall cupboards in front of the window.

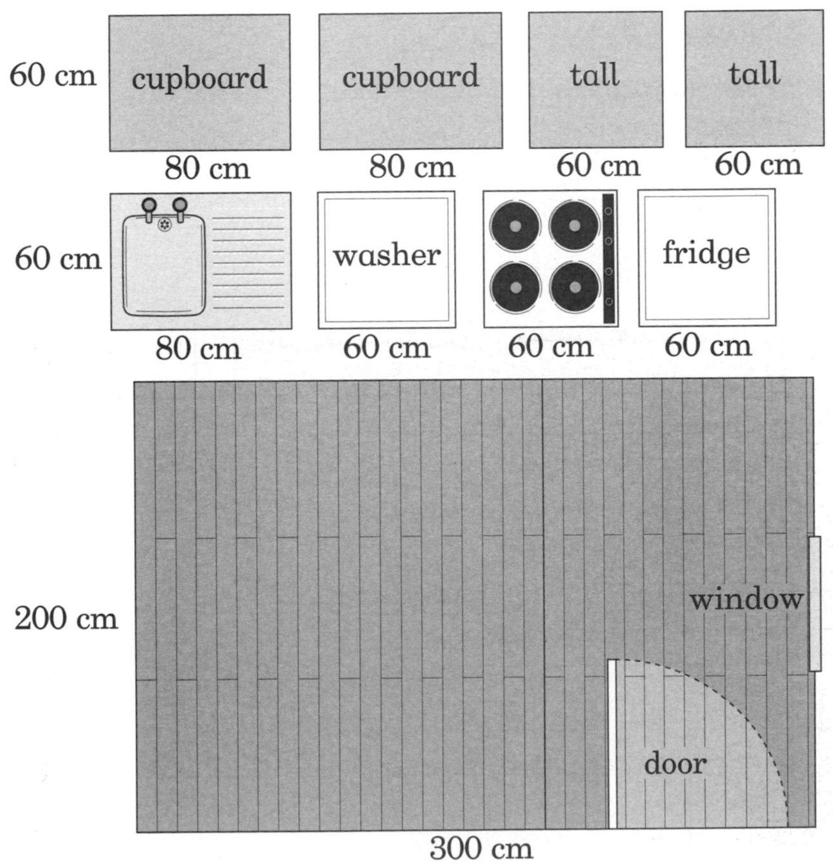

3 Finding lengths

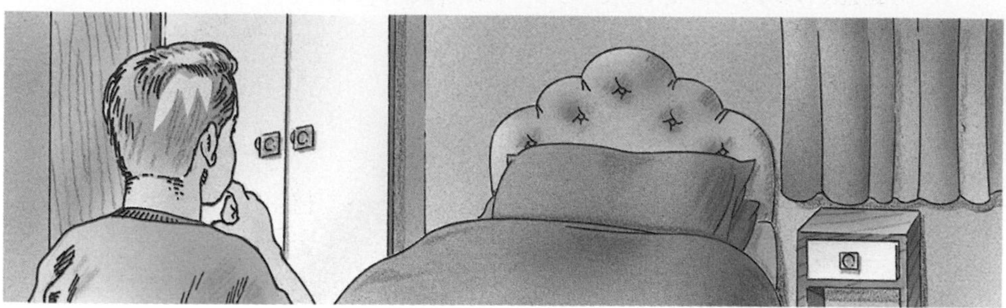

Jonathon is planning a new bedroom.

He has measured his bedroom.
It measures 300 cm by 360 cm.
He uses a scale of 20 cm in his bedroom to be 1 cm on his plan.
To calculate the size of his plan he works out

$300\text{ cm} \div 20 = 15\text{ cm}$

He draws one side of his bedroom on squared paper.

He calculates the other side

$360\text{ cm} \div 20 = 18\text{ cm}$

He draws the other side.

He puts the door in the correct place like this.

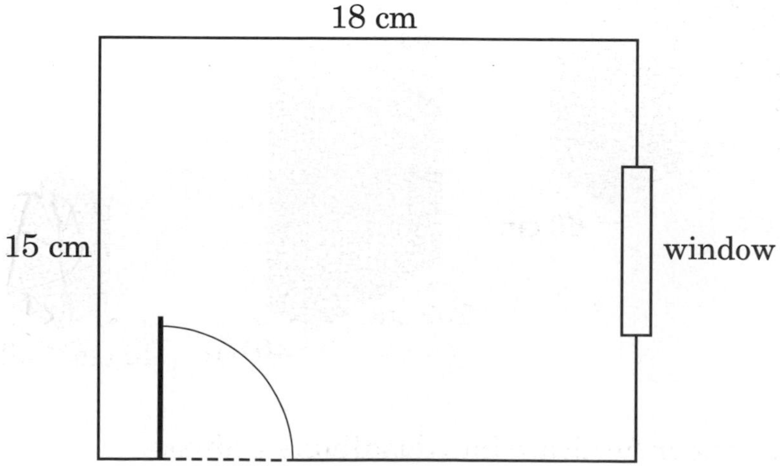

Exercise 12:12

This is the furniture Jonathon wants in his bedroom.
Make scale models of the pieces of furniture on squared paper.
Remember to divide each length by 20.

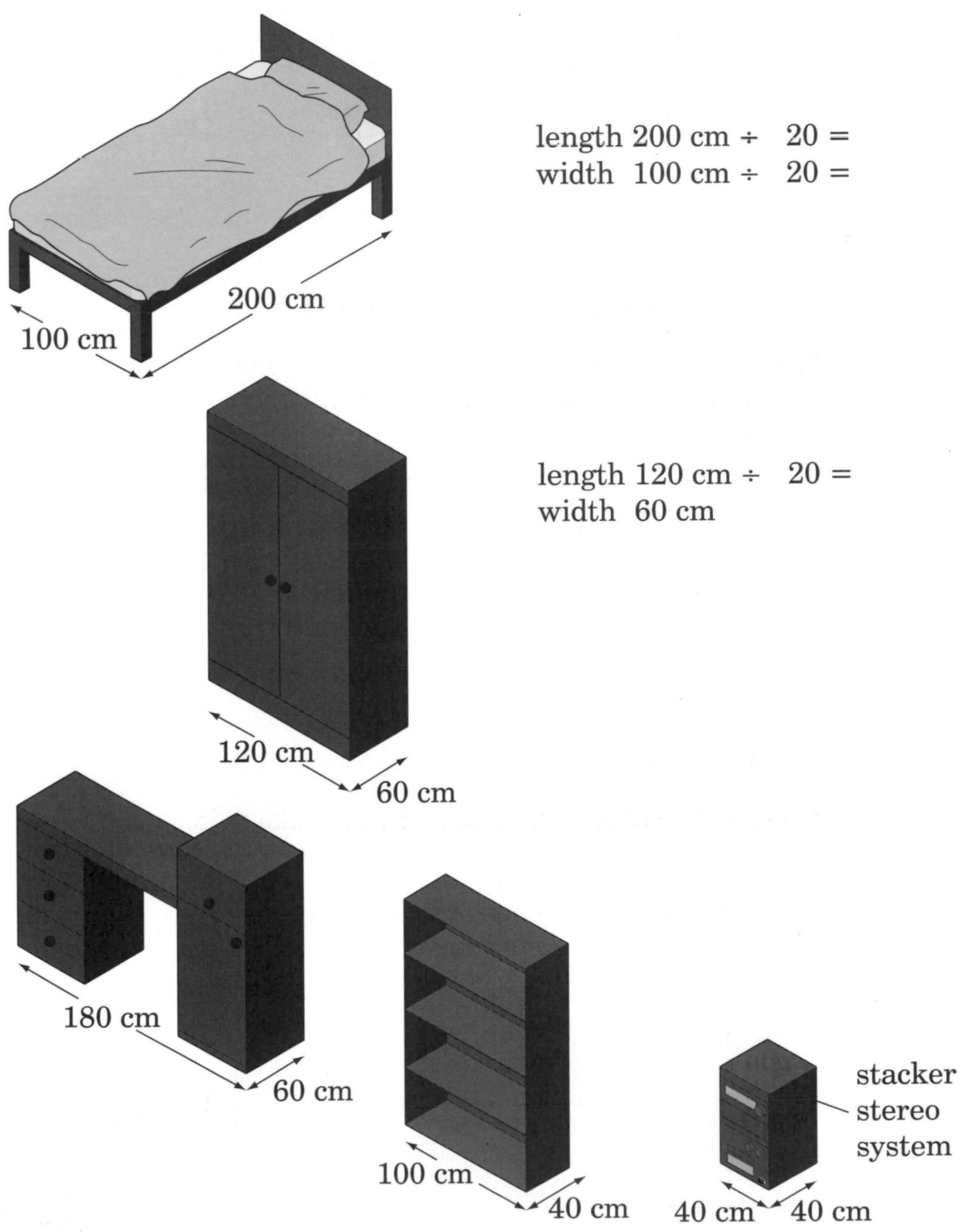

length 200 cm ÷ 20 =
width 100 cm ÷ 20 =

length 120 cm ÷ 20 =
width 60 cm

G 1, 2 Now put your furniture into Jonathon's bedroom.

1 Which of these units, millimetres (mm), centimetres (cm), metres (m) or kilometres (km), would you use to measure:

a the length of your arm?

b the width of your desk?

c the length of a swimming pool?

d the length of a baby's little toe?

e the distance from London to Bristol?

f the length of a ladybird?

g the distance from the Earth to the Moon?

h the length of a spider's leg?

2 *Estimate* the lengths of these insects in centimetres (cm). This ruler is marked in centimetres. Use it to help you.

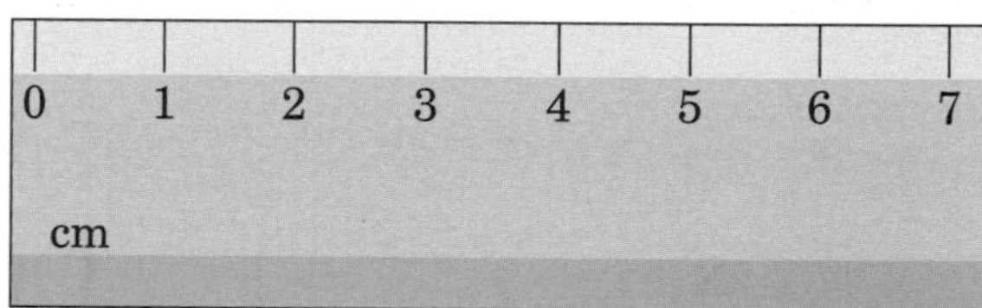

a 

b 

c 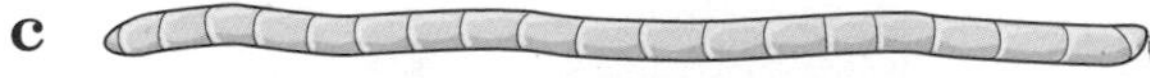

d 

Now use a ruler to measure the lengths of the insects exactly in centimetres and millimetres.

3 Measure these lines.
Write down their lengths in centimetres and millimetres.

a 

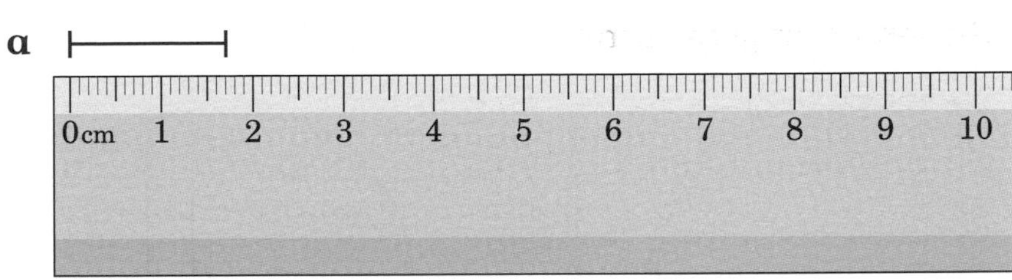

b

c

d

4 Measure these lines and write down their lengths in millimetres.

a

b

c

d

e

5 Jane's classroom measures 600 cm by 800 cm.
She uses a scale of 20 cm in her classroom to 1 cm on her plan.

a Calculate the length of the classroom.

$600 \div 20 =$

b Calculate the width of the classroom.

c The teacher's desk measures 160 cm by 80 cm.
How big will the teacher's desk measure on the plan?

d Jane's desk measures 80 cm by 80 cm.
How big will Jane's desk be on the plan?

Time 2

Write the times shown on these clocks in words.

1

2

3

4

5

6

7

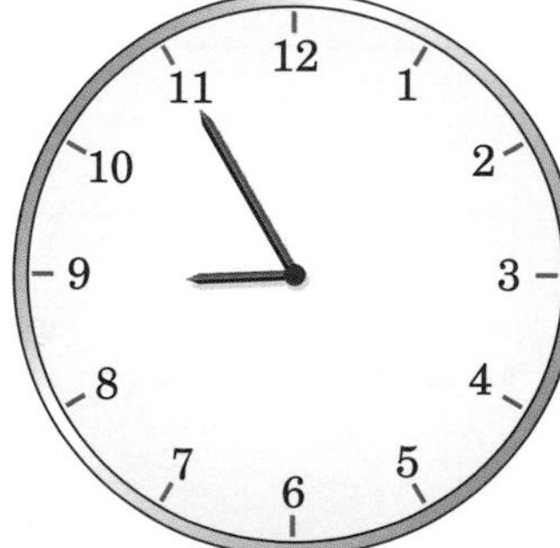

8

13 Algebra: into the unknown

CORE

1 **Inverse operations**

2 **Equations**

3 **Trial and improvement**

QUESTIONS

HELP YOURSELF **Multiplication 3**

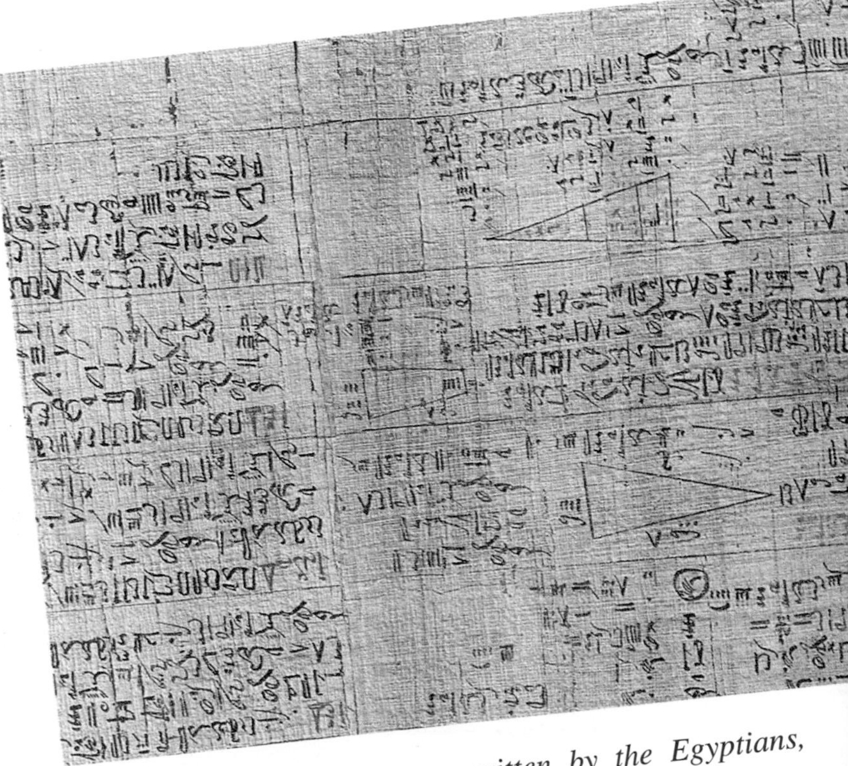

The earliest equations were written by the Egyptians, between 1650 and 1550 BC.

1 Inverse operations

Christopher has knitted a scarf.

His puppy undoes the scarf.

Christopher is back to where he started.
'Undoing the scarf' is called the **inverse** of 'knitting the scarf'.

Inverse An **inverse** is the action that returns you to where you started.

Example

Put on your shoes is the **inverse** of **Take off your shoes**

Exercise 13:1

1 Write down the inverse of:

a open the door
b walk up the stairs
c turn left
d walk backwards
e open the window
f turn on the light
g put on your coat
h get dressed
i turn on the TV
j put on your hat

Finding the inverse in mathematics

Example

Start with some blocks:

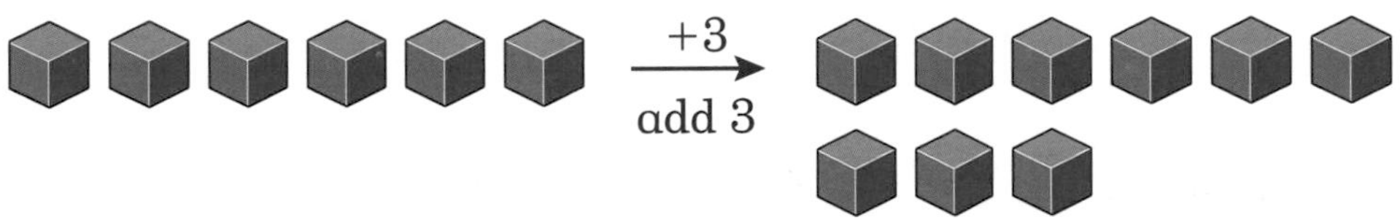

To get back to where you started:

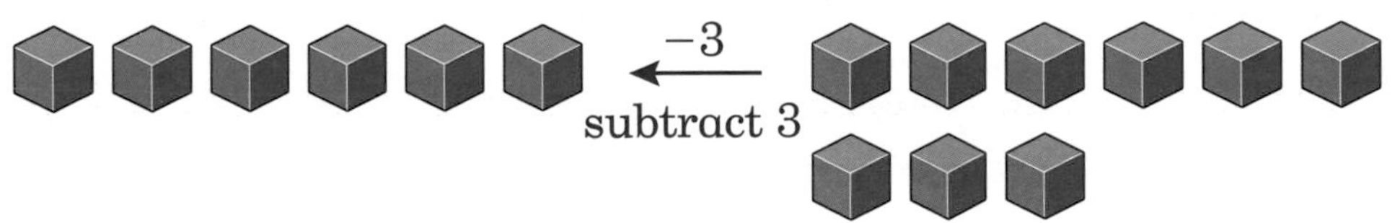

The inverse of +3 is −3
add 3 subtract 3

Exercise 13:2

1 Write down the inverse of these.
You can use blocks to help you.

a Add 2
b Add 4
c Add 8
d +5
e +7
f +12

Example

Start with some blocks:

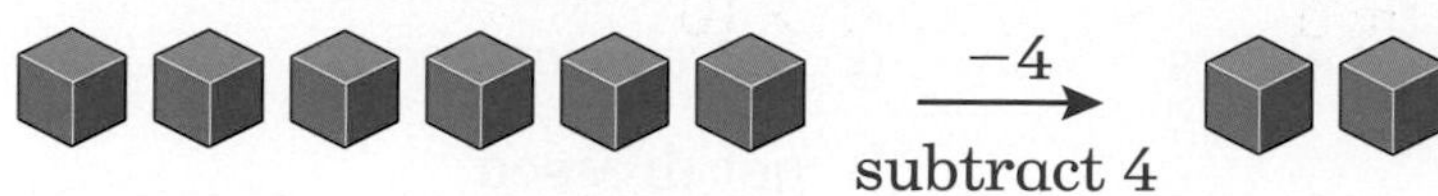

To get back to where you started:

The inverse of −4 is +4
subtract 4 add 4

2 Write down the inverse of these.
You can use blocks to help you.

a Subtract 3
b Subtract 7
c Subtract 10
d −2
e −6
f −11

Example

Start with some blocks:

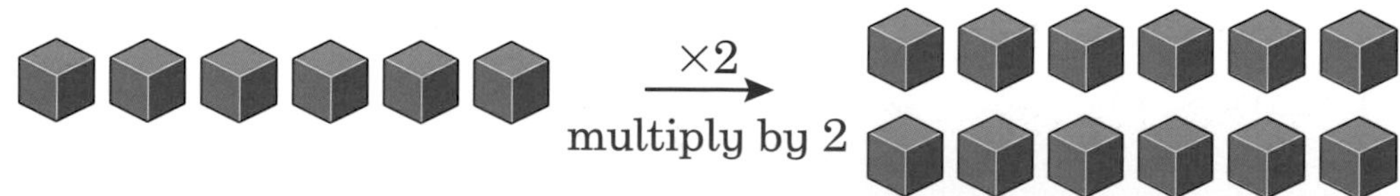

To get back to where you started:

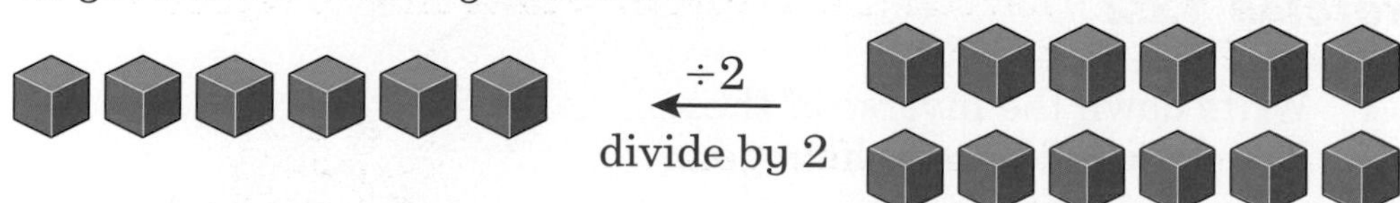

The inverse of ×2 is ÷2
multiply by 2 divide by 2

3 Write down the inverse of these.
You can use blocks to help you.

a Multiply by 2 **c** $\times 3$

b Multiply by 5 **d** $\times 10$

Example

Start with some blocks:

$\div 3$ → divide by 3

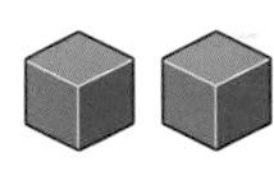

To get back to where you started:

$\times 3$ ← multiply by 3

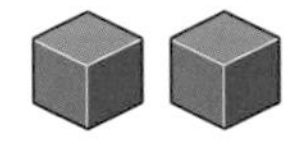

The inverse of $\div 3$ divide by 3 is $\times 3$ multiply by 3

---

4 Write down the inverse of these.
You can use blocks to help you.

a Divide by 5 **c** $\div 10$

G 1, 2 **b** Divide by 3 **d** $\div 2$

Function machine You can use a **function machine** to represent 'add 10'.

Example Put 8 into the function machine.
You get the answer 18.

Exercise 13:4

1 Philip's puppy has eaten some chocolate bars.
There are 2 left.
There were 8.
Write down the equation.
Use c for the number of chocolate bars the puppy has eaten.

2 Jo's guinea pigs have eaten some carrots.
There are 7 left.
There were 11.
Write down the equation.
Use c for the number of carrots the guinea pigs have eaten.

3 The teachers have eaten some biscuits.
There are 8 left.
There were 15.
Write down the equation.
Use b for the number of biscuits the teachers have eaten.

4 The children have eaten some sweets.
There is 1 left.
There were 12.
Write down the equation.
Use s for the number of sweets the children have eaten.

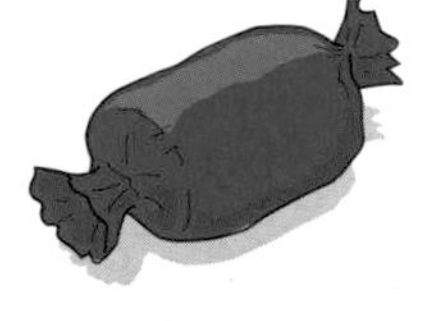

5 The nurse has used some plasters.
There are 5 left.
There were 9.
Write down the equation.
Use p for the number of plasters the nurse has used.

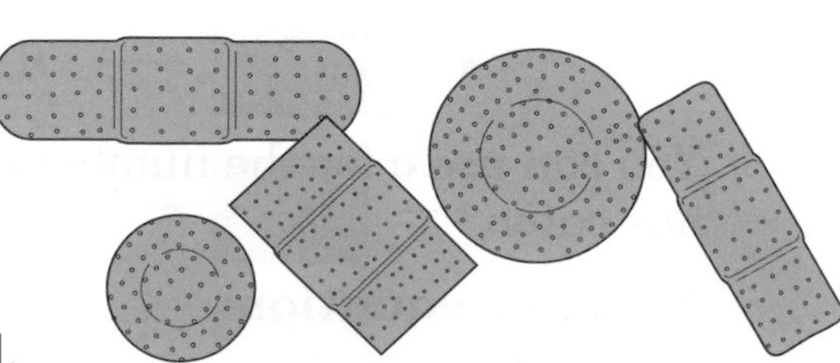

6 Henrietta spent some money.
She has £4 left.
She had £20.
Write down the equation.
Use m for the amount of money Henrietta has spent.

Example You can find the answers to equations using robots.

Write the equation:

Work it backwards using the inverse:

The answer is $n = 7$

Exercise 13:5

Use this method to solve these equations.

1 $e + 4 = 7$

$e \longrightarrow$ [+ 4] $\longrightarrow 7$

[ ] $\longleftarrow$ [− 4] $\longleftarrow 7$

The answer is $e =$ [ ]

2 $x + 3 = 8$

$x \longrightarrow$ [+ 3] $\longrightarrow 8$

[ ] $\longleftarrow$ [− 3] $\longleftarrow 8$

The answer is $x =$ [ ]

3 $t + 5 = 9$

$t \longrightarrow$ [$+5$] $\longrightarrow$ 9

[ ] $\longleftarrow$ [ ] $\longleftarrow$ [ ]

The answer is $t =$ [ ]

4 $y + 4 = 8$

$y \longrightarrow$ [$+4$] $\longrightarrow$ [ ]

[ ] [ ] [ ]

The answer is $y =$ [ ]

G 3, 4, 5, 6

Example Inverses can be used to solve other equations.

$t - 5 = 12$

$t \longrightarrow$ [-5] $\longrightarrow 12$

$17 \longleftarrow$ [$+5$] $\longleftarrow 12$

The answer is $t = 17$

Use this method to solve these equations.

5 $x - 4 = 8$

$x \longrightarrow$ [-4] $\longrightarrow 8$

[ ] $\longleftarrow$ [ ] $\longleftarrow 8$

The answer is $x =$ [ ]

7 $n - 5 = 10$

n 

[ ] 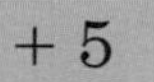 [$+5$] [ ]

The answer is $n =$ [ ]

6 $g - 3 = 12$

$g \longrightarrow$ [ ] $\longrightarrow$ [ ]

[ ] $\longleftarrow$ [ ] $\longleftarrow 12$

The answer is $g =$ [ ]

8 $p - 7 = 10$

p 

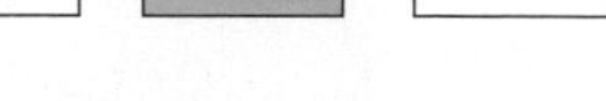

The answer is $p =$ [ ]

G 4, 5, 6, 7

Example Solve the equation $a \times 3 = 12$

$a \longrightarrow$ [$\times 3$] $\longrightarrow 12$

[ ] $\longleftarrow$ [$\div 3$] $\longleftarrow 12$

The answer is $a = 4$

Exercise 13:6

Solve these equations using inverse function machines.

1 $p \times 4 = 12$

$p \longrightarrow$ [$\times 4$] $\longrightarrow 12$

[ ] $\longleftarrow$ [$\div 4$] $\longleftarrow$ [ ]

The answer is $p =$ [ ]

2 $y \times 5 = 30$

$y \longrightarrow$ [$\times 5$] $\longrightarrow$ [ ]

[ ] $\longleftarrow$ [ ] $\longleftarrow 30$

The answer is $y =$ [ ]

3 $h \times 3 = 12$

h

The answer is $h =$ [ ]

4 $e \times 10 = 70$

y [$\times$ ...]

The answer is $e =$ [ ]

Example Solve the equation $x \div 2 = 7$

$x \longrightarrow$ [$\div 2$] $\longrightarrow 7$

$14 \longleftarrow$ [$\times 2$] $\longleftarrow 7$

The answer is $x = 14$

Solve these equations using inverse function machines.

5 $y \div 3 = 4$

$y \rightarrow$ [$\div 3$] $\rightarrow 4$

[ ] $\leftarrow$ [$\times 3$] $\leftarrow 4$

The answer is $y =$ [ ]

6 $p \div 5 = 3$

$p \rightarrow$ [$\div 5$] $\rightarrow 3$

[ ] [$\times 5$] [ ]

The answer is $p =$ [ ]

7 $k \div 4 = 10$

k [ ] [ ]

[ ] [ ] [ ]

The answer is $k =$ [ ]

8 $c \div 2 = 8$

c [ ] [ ]

[ ] [ ] [ ]

The answer is $c =$ [ ]

Example An equation can be given in words.
You can change the words into algebra.

I think of a number.	I call the number n
I add 5.	$n + 5$
The answer is 12.	$n + 5 = 12$

Use an inverse function machine to find out what the number is.

$n + 5 = 12$

$n \rightarrow$ [$+ 5$] $\rightarrow 12$

[ ] $\leftarrow$ [$- 5$] $\leftarrow 12$

The answer is $n = 7$

The number I thought of was 7.

Exercise 13:7

Use this method to find these numbers.

1 I think of a number and add 3. The answer is 12.

$n + 3 = 12$

$n \longrightarrow$ [+ 3] $\longrightarrow 12$

[ ] $\longleftarrow$ [− 3] $\longleftarrow$ [ ]

The answer is $n =$ [ ]

The number I thought of was ....

2 I think of a number and add 7. The answer is 13.

$n + 7 = 13$

n [ ] 13

[ ] [ ] [ ]

The answer is $n =$ [ ]

The number I thought of was ....

3 I think of a number and add 8. The answer is 17.

n + [ ] = [ ]

n [ ] [ ]

[ ] [ ] [ ]

The answer is $n =$ [ ]

The number I thought of was ....

4 I think of a number and add 10. The answer is 20.
Find the number I thought of.

5 I think of a number and subtract 4. The answer is 8.

$n - 4 = 8$

n | -4 | 8

☐ | ☐ | 8

The answer is $n =$ ☐

The number I thought of was ....

6 I think of a number and subtract 6. The answer is 11.

$n -$ ☐ $=$ ☐

n | ☐ | ☐

☐ | ☐ | ☐

The answer is $n =$ ☐

The number I thought of was ....

7 I think of a number and subtract 10. The answer is 12.
Find the number I thought of.

8 I think of a number and subtract 3. The answer is 29.
Find the number I thought of.

3 Trial and improvement

Aisha is unhappy.

She can't solve this equation.

Aisha decides to solve her equation by guessing.

She has to solve $x + 24 = 120$

Aisha's first guess is 80.	$80 + 24 = 104$	too small
Aisha's second guess is 100.	$100 + 24 = 124$	too big
Aisha tries 95.	$95 + 24 = 119$	too small
Aisha tries 96.	$96 + 24 = 120$	correct

The answer is $x = 96$

This is called trial and improvement.

Aisha is happy.

She wants to try another equation.

We often use a calculator to help us solve equations using trial and improvement.

Example Solve $n + 19 = 82$

Aisha tries different numbers.
She writes her results in a table.

Value of n	Value of $n + 19$	
60	$60 + 19 = 79$	too small
70	$70 + 19 = 89$	too big
63	$63 + 19 = 82$	✓

The answer is $n = 63$

Exercise 13:8

H 2 Solve these equations using trial and improvement.
Put your work in a table.
Keep trying until you get the correct answer.

1 $m + 14 = 71$

Value of m	Value of $m + 14$	

2 $y + 21 = 102$

Value of y	Value of $y + 21$	

3 $k + 79 = 111$

Value of k	Value of $k + 79$	

4 $x + 85 = 142$

Value of x	Value of $x + 85$	

Example You can use trial and improvement to solve any equation.

Solve $p - 12 = 43$

Value of p	Value of $p - 12$	
60	$60 - 12 = 48$	too big
50	$50 - 12 = 38$	too small
56	$56 - 12 = 44$	too big
55	$55 - 12 = 43$	✓

The answer is $p = 55$

 2 Use trial and improvement to solve these equations.

5 $x - 22 = 50$

Value of x	Value of $x - 22$

6 $n - 19 = 78$

Value of n	Value of $n - 19$

7 $w - 47 = 35$

Value of w	Value of $w - 47$

G 8, 9 **8** $h - 84 = 117$

Value of h	Value of $h - 84$

1 Write down the inverse of these.

a Put on your socks.

b Walk down the stairs.

c Turn off the light.

d Take off your coat.

e Jump up.

f Shut the curtains.

2 Write down the inverse of these.

a Add 3

b Add 10

c +6

d +15

3 Write down the inverse of these.

a Subtract 2

b Subtract 9

c −8

d −1

4 Write down the inverse of these.

a Multiply by 4

b Multiply by 6

c ×7

d ×2

5 Write down the inverse of these.

a Divide by 6

b Divide by 8

c ÷3

d ÷5

6 Draw the inverse function machines.
Make sure the arrows go the correct way.

a → [+2] →

b → [+6] →

c → [+8] →

d → [−2] →

e → [−10] →

f → [−7] →

g → [×4] →

h → [×7] →

i → [×8] →

j → [÷7] →

k → [÷5] →

l → [÷9] →

7 Mary's puppy has eaten some chocolate bars.
There are 6 left.
There were 12.
Write down the equation.
Use c for the number of chocolate bars the puppy has eaten.

8 Geoff has eaten some sweets.
There are 5 left.
There were 9.
Write down the equation.
Use s for the number of sweets Geoff has eaten.

9 Solve these equations.

a $s + 5 = 10$

[ ] ← [−5] ← 10

b $x + 2 = 7$

[ ] ← [ ] ← [ ]

c $a - 4 = 11$

[ ] ← [ ] ← [ ]

d $t - 3 = 12$

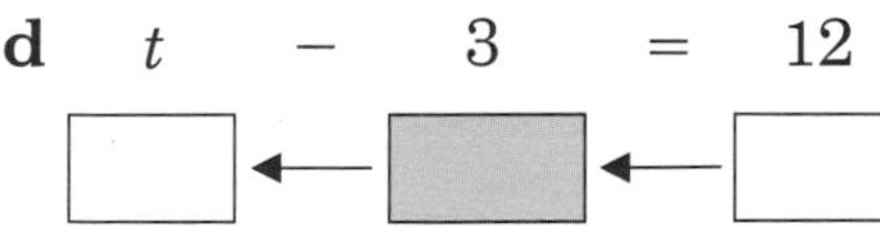

e $d \times 2 = 8$

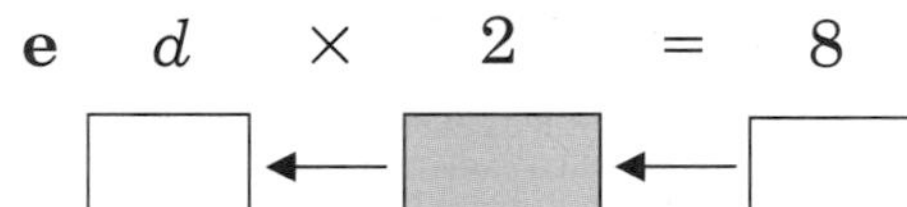

f $e \times 5 = 50$

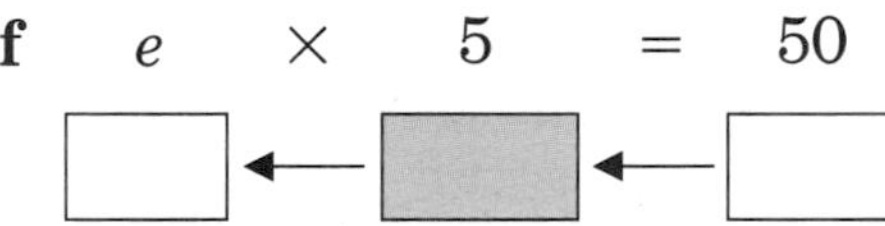

g $f \div 2 = 5$

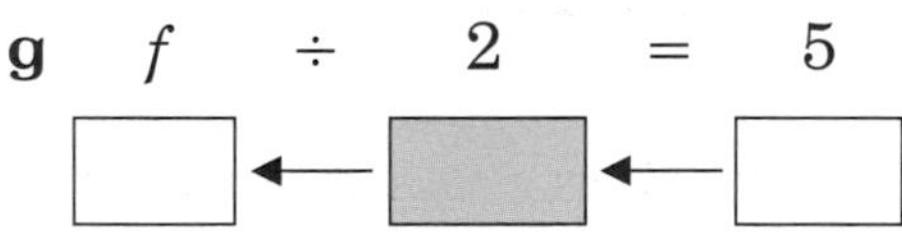

h $g \div 10 = 3$

[ ] ← [ ] ← [ ]

10 I think of a number and add 5. The answer is 16.
What is the number I thought of?

11 I think of a number and subtract 2. The answer is 17.
What is the number I thought of?

2

12 Use trial and improvement to solve these equations.

a $s + 24 = 93$

b $v - 64 = 107$

c $h + 125 = 231$

d $l - 28 = 97$

Multiplication 3

Copy these into your book.
Work out the answers.

Exercise 1	Exercise 2	Exercise 3	Exercise 4
1 3×2	**1** 3×4	**1** 6×8	**1** 2×10
2 6×4	**2** 3×8	**2** 5×3	**2** 6×10
3 3×5	**3** 7×3	**3** 6×9	**3** 10×10
4 8×2	**4** 5×2	**4** 7×2	**4** 9×10
5 4×3	**5** 6×3	**5** 5×4	**5** 3×100
6 4×5	**6** 3×9	**6** 2×9	**6** 3×10
7 4×2	**7** 4×4	**7** 6×10	**7** 5×100
8 6×5	**8** 6×2	**8** 5×5	**8** 8×100
9 5×2	**9** 8×3	**9** 6×2	**9** 7×10
10 2×4	**10** 3×5	**10** 5×6	**10** 6×100
11 9×2	**11** 4×8	**11** 4×3	**11** 4×10
12 4×7	**12** 7×2	**12** 7×4	**12** 9×100
13 3×6	**13** 3×7	**13** 5×9	**13** 5×10
14 9×3	**14** 8×2	**14** 5×8	**14** 8×10
15 4×6	**15** 4×9	**15** 4×4	**15** 4×100
16 6×6	**16** 6×7	**16** 5×7	**16** 7×100

14 Area and perimeter

CORE

1 **Shapes**

2 **Areas of rectangles**

3 **Perimeters of shapes**

QUESTIONS

HELP YOURSELF **Division 3**

Four countries make up the United Kingdom. You may be surprised to see the area they cover compared with the number of people who live in each country:

- England has an area of 50 333 square miles and a population of 48.3 million
- Scotland has an area of 30 405 square miles and a population of 4.9 million
- Wales has an area of 8016 square miles and a population of 2.9 million
- Northern Ireland has an area of 5462 square miles and a population of 1.6 million.

But compare all these with Canada, which has an area of 3 851 800 square miles and a population of only 27.7 million.

1 Shapes

Carpet-fitters are covering the floor.

The carpet tiles are squares.

Area The **area** is how much space something covers.

Example Sarah counts the area by putting a number in each square.

Each square measures 1 cm by 1 cm.
You write this: $1\,\text{cm}^2$
There are 6 squares.

The area is $6\,\text{cm}^2$.

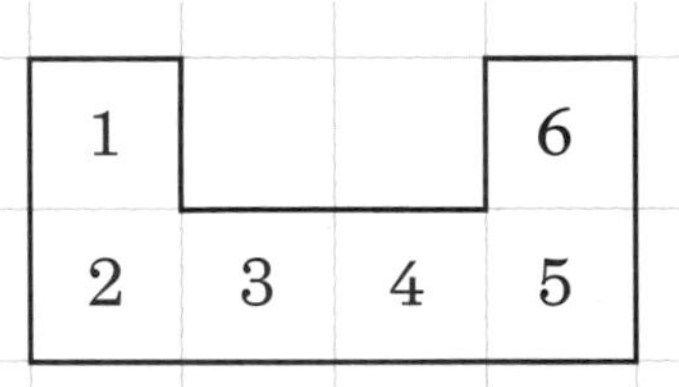

Exercise 14:1

H 1 Find the area of these shapes by counting squares.

1

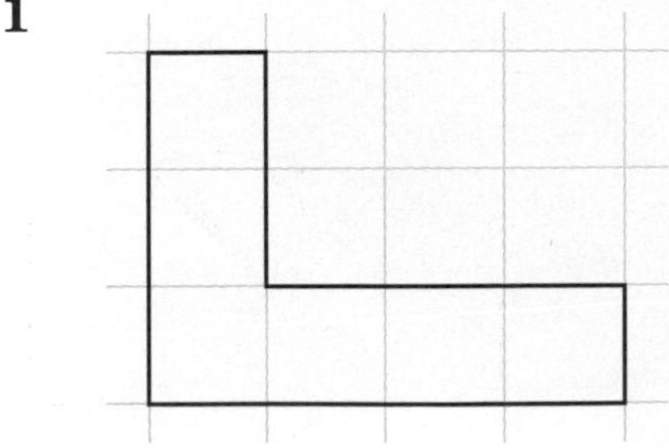

2

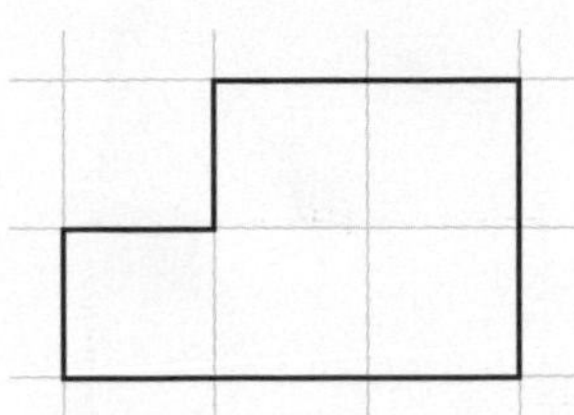

3

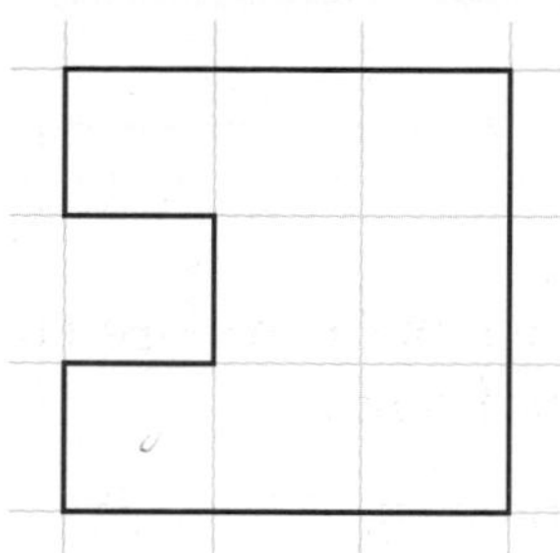

4

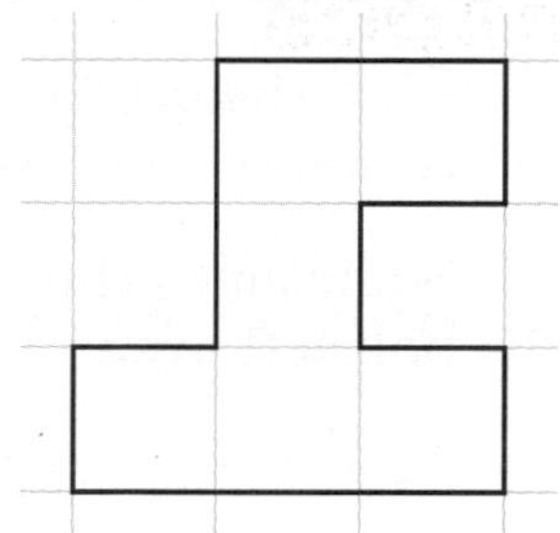

Example Find the area of this shape.

You can add two half squares.

Two $\frac{1}{2}$ squares make 1 whole square.

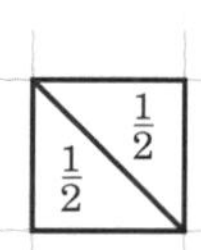

$\frac{1}{2}$ 1 2 3

$\frac{1}{2}$ 4 5

6 7

Area = 8 square centimetres = 8 cm^2

Exercise 14:2

H 2 Find the area of these shapes.

1

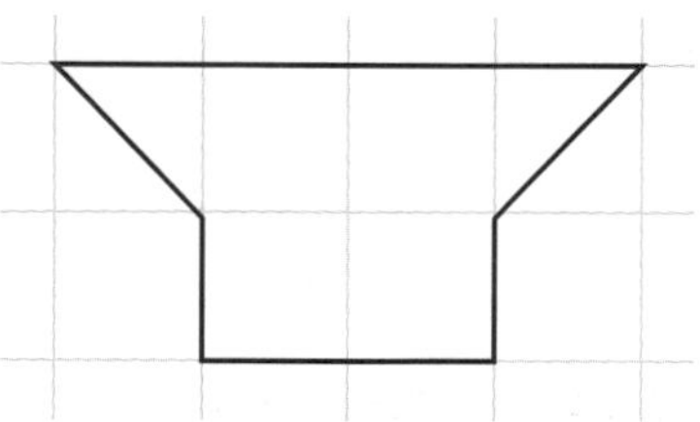

3

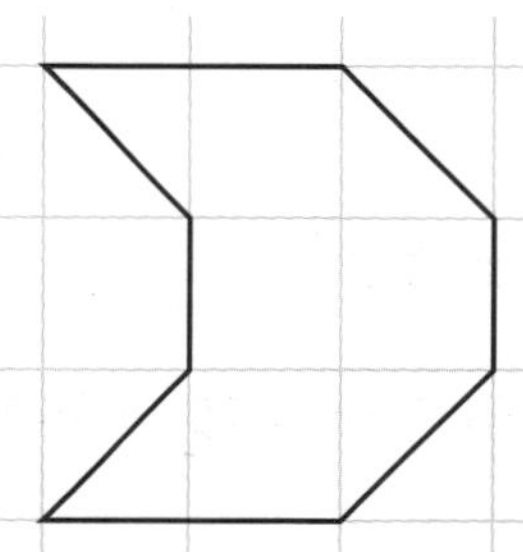

2

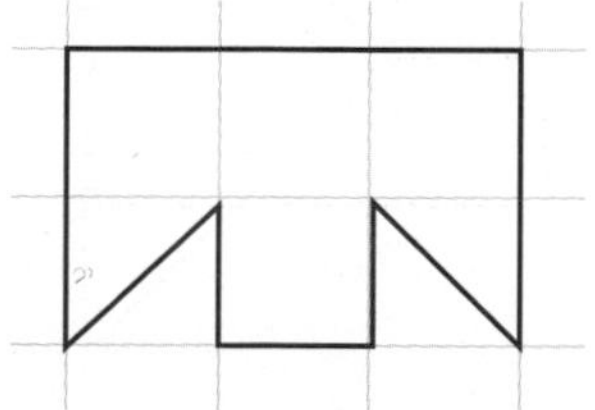

4

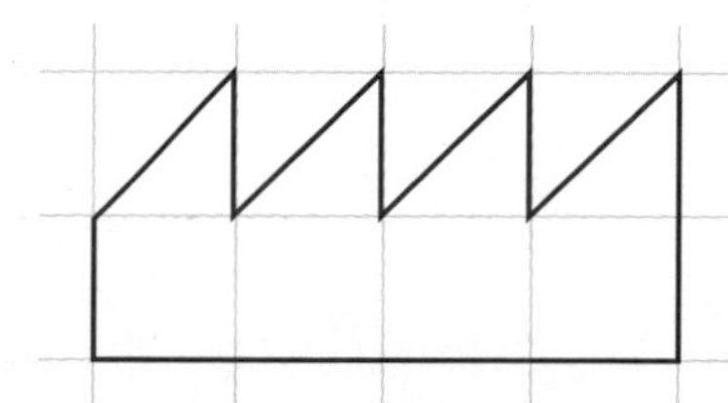

Example Find the area of this shape.

Sometimes there is half left over.

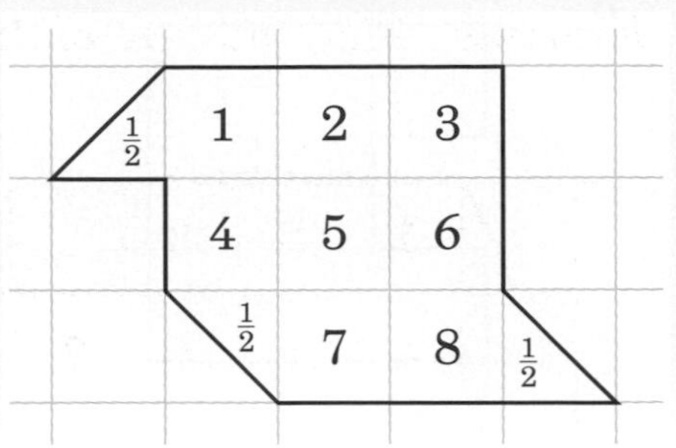

Area $= 8 + \frac{1}{2} + \frac{1}{2} + \frac{1}{2}$

$= 8 + 1 + \frac{1}{2}$

$= 9\frac{1}{2}$ square centimetres $= 9\frac{1}{2}$ cm^2

Exercise 14:3

H 3 Find the area of these shapes.

1

3

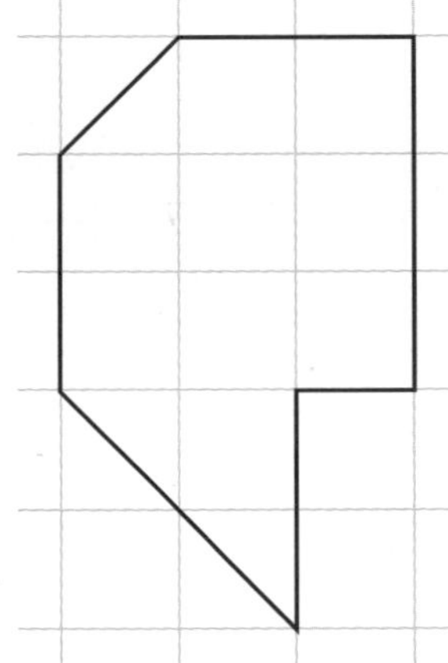

2

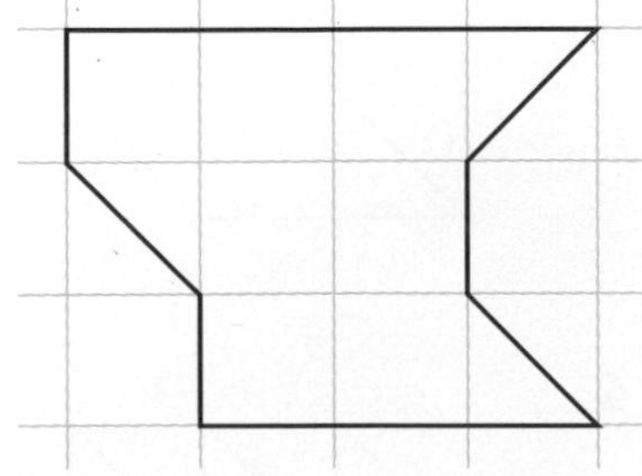

4

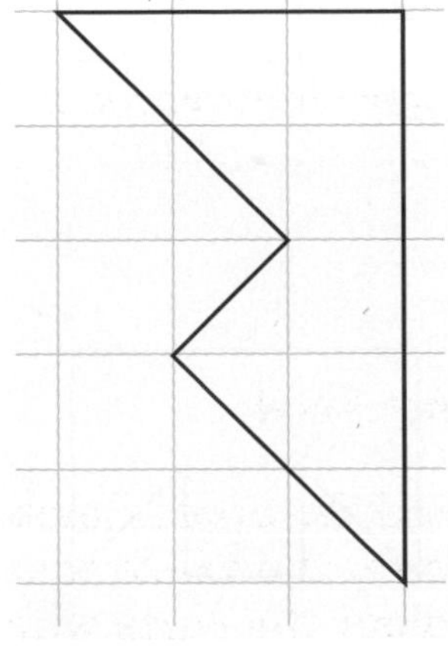

Example Find the area of this shape.

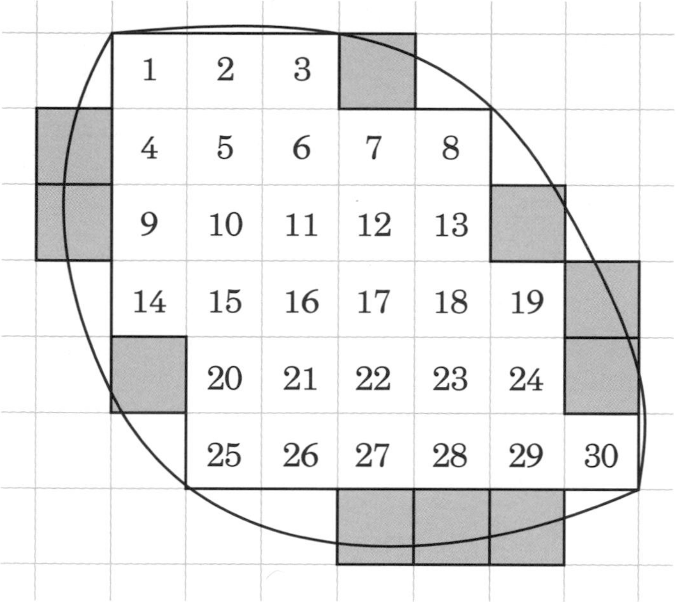

	1	2	3			
	4	5	6	7	8	
	9	10	11	12	13	
	14	15	16	17	18	19
		20	21	22	23	24
		25	26	27	28	29	30

Not all shapes have straight sides.

First count the whole squares.
There are 30.

Then colour in the parts which are more than half a square.

There are 10. (Ignore the parts which are less than half a square.)

You can't find the exact area.

Whole squares + parts more than half a square = approximate area

30 squares + 10 squares = 40 squares

The approximate area is 40 cm^2.

Exercise 14:4

W 1 Find the approximate area of these shapes.
Count the whole squares.
Colour the parts which are more than half a square.

1

Whole squares + parts more than half a square = approximate area

.... + ... = ...

The approximate area is .... cm^2

2

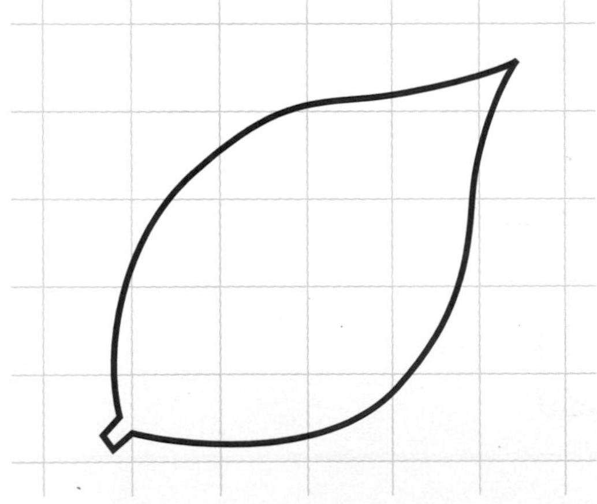

Whole squares + parts more than half a square = approximate area

.... + ... = ...

The approximate area is .... cm^2

3

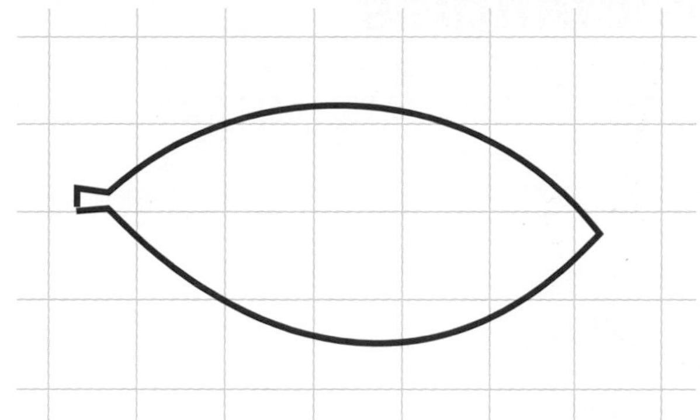

Whole squares + parts more than half a square = approximate area

.... + ... = ...

The approximate area is .... cm^2

4

Whole squares + parts more than half a square = approximate area

.... + ... = ...

The approximate area is .... cm^2

2 Areas of rectangles

Julian and Justin have laid the carpet tiles.

Example This is a drawing of a rectangle.

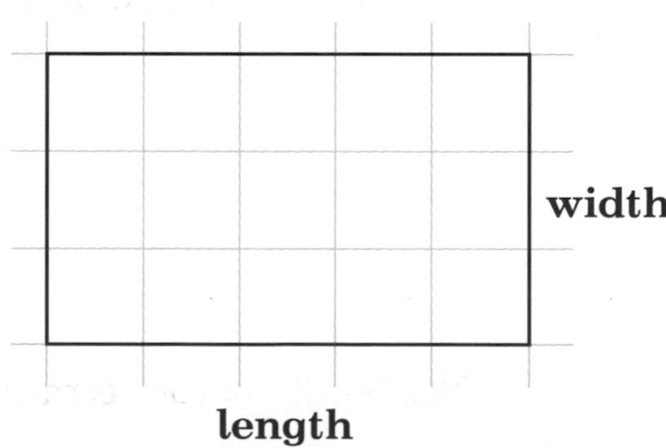

The **length** of the rectangle is 5 cm.

The **width** of the rectangle is 3 cm.

There are 3 rows of 5.
The area of the rectangle is $3 \times 5 = 15$ squares.

The area of the rectangle is 15 cm^2.

Exercise 14:5

2 Use the worksheet to find the area of these rectangles.

1

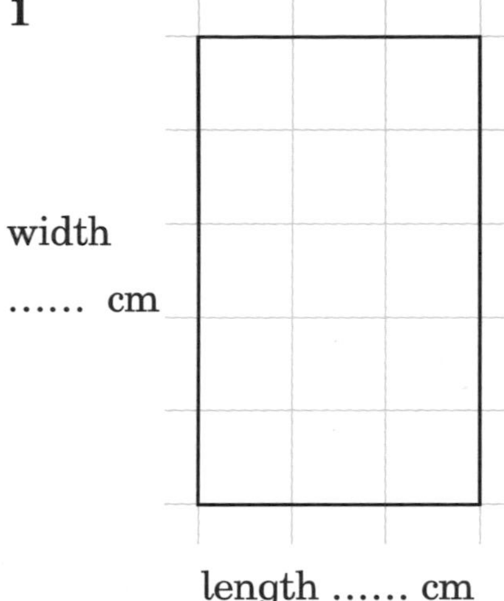

The length of the rectangle is ...... cm

The width of the rectangle is ...... cm

The area of the rectangle is ...... cm^2

2

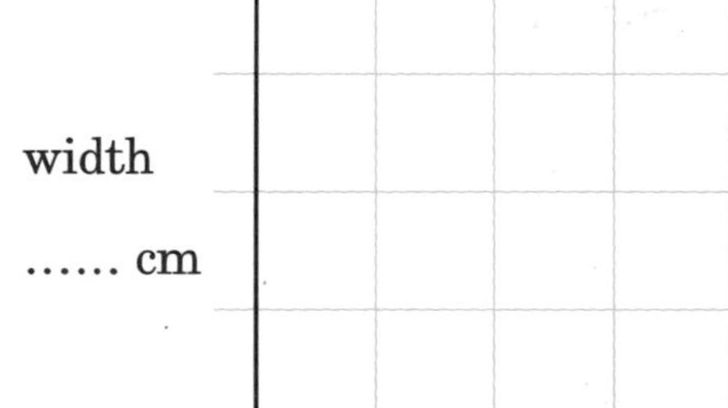

The length of the rectangle is ...... cm

The width of the rectangle is ...... cm

The area of the rectangle is ...... cm^2

3

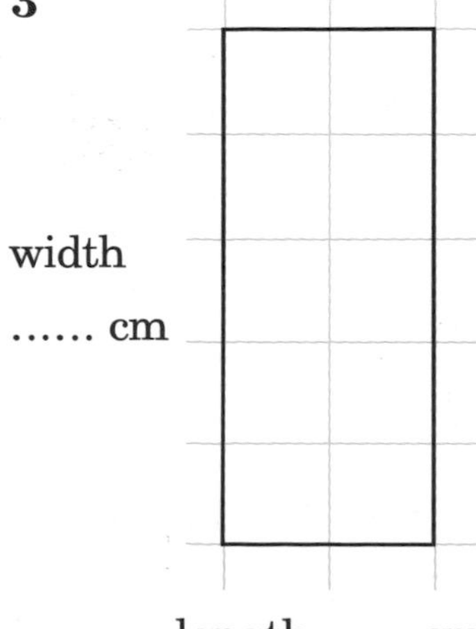

The length of the rectangle is ...... cm

The width of the rectangle is ...... cm

The area of the rectangle is ...... cm^2

4

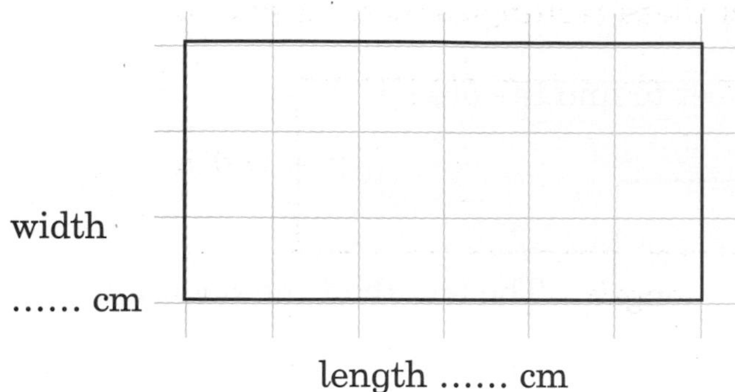

The length of the rectangle is ...... cm

The width of the rectangle is ...... cm

The area of the rectangle is ...... cm^2

Area of a rectangle

Area of a rectangle = ***l***ength × ***w***idth

$\boldsymbol{A = l \times w}$ or $\boldsymbol{A = lw}$

Example Find the area of this rectangle by multiplying.

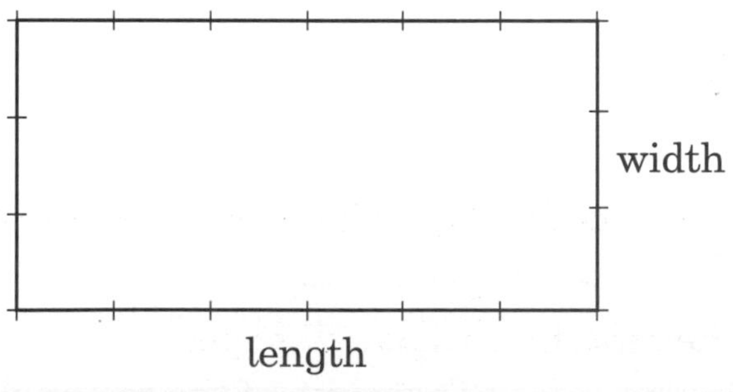

Length = 6 cm

Width = 3 cm

Area = $l \times w$
Area = 6 cm × 3 cm
Area = 18 cm^2

W 3 Find the area of these rectangles by multiplying.

5

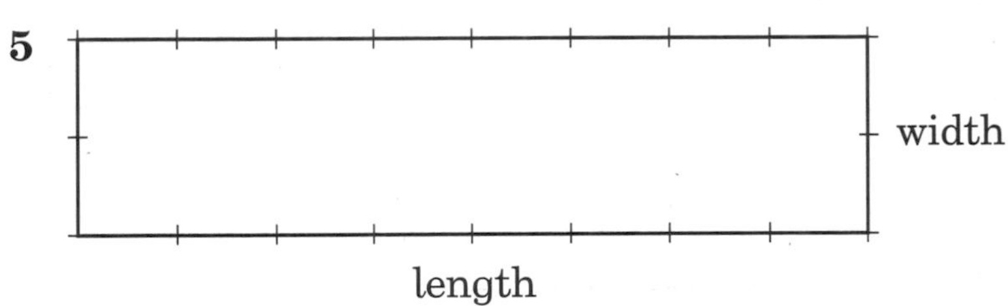

Length = …… cm

Width = …… cm

Area = …… cm × …… cm

Area = …… cm^2

6

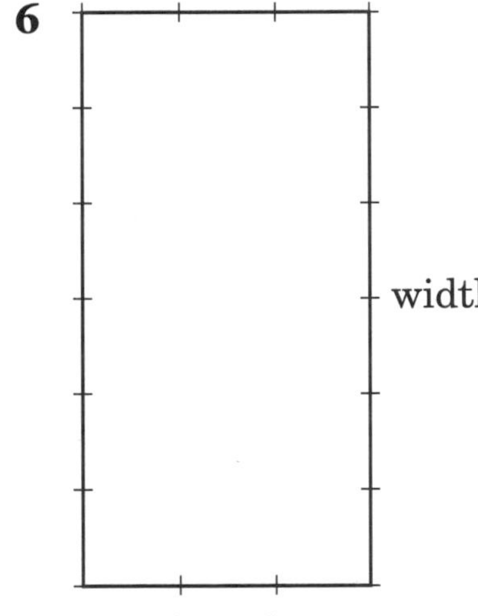

Length = …… cm

Width = …… cm

Area = …… cm × …… cm

Area = …… cm^2

Exercise 14:6 To find all the pentominoes

You need some centimetre squared paper.

A **pentomino** is a shape made from 5 squares.

Cut out 5 squares.

This **is** a pentomino: 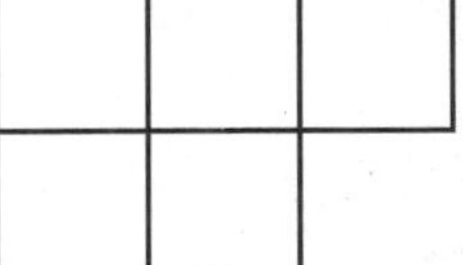

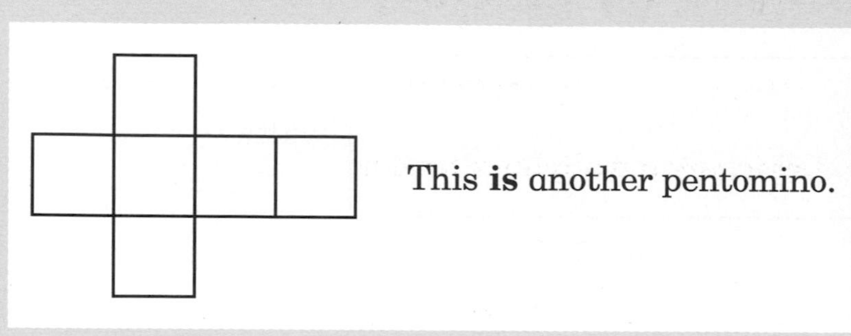

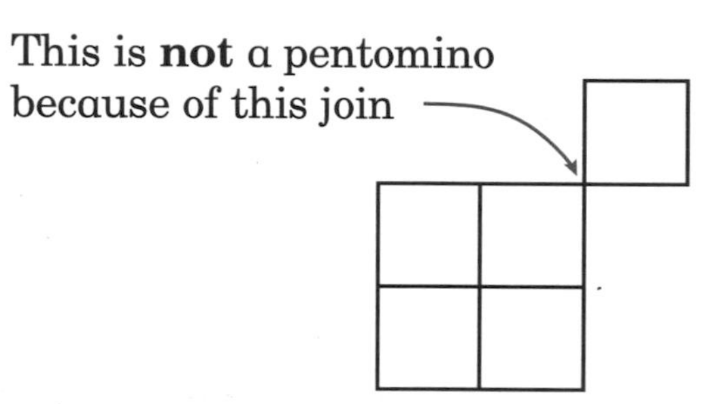

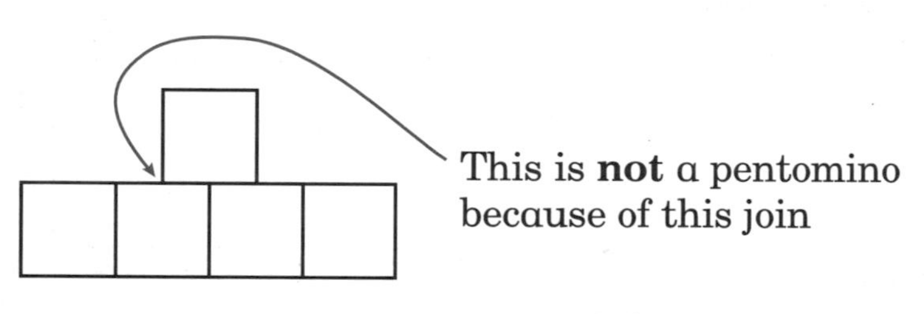

1. Arrange 5 squares into a pentomino.

2. Copy it on to paper.

3. Try to find as many different pentominoes as you can.

   Finding 8 different pentominoes is **good**.
   Finding 10 different pentominoes is **very good**.
   Finding 12 different pentominoes is **excellent**.

4. Make a poster of your pentominoes.

5. Write down how you made sure that you could get as many as possible.

3 Perimeters of shapes

Angela is sticking up a wallpaper border.

Perimeter The **perimeter** is the distance all the way round.

Example Find the perimeter of this rectangle.

Angela writes numbers on the edge of the shape.

She doesn't want to miss any. The dot reminds her where she started.

The perimeter is 14 cm.

Exercise 14:7

Find the perimeter of these shapes.

1

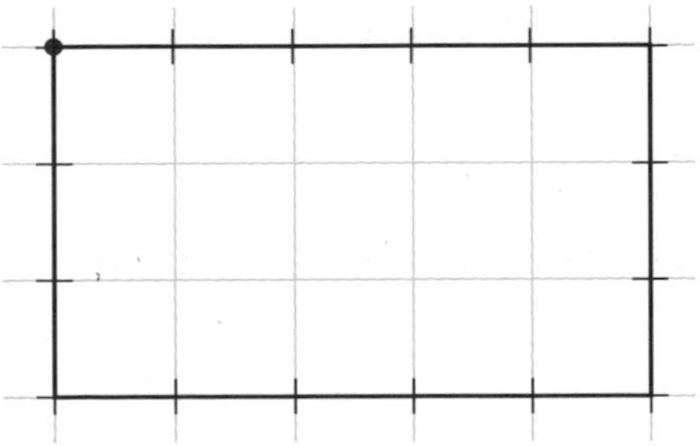

The perimeter is …… cm

2

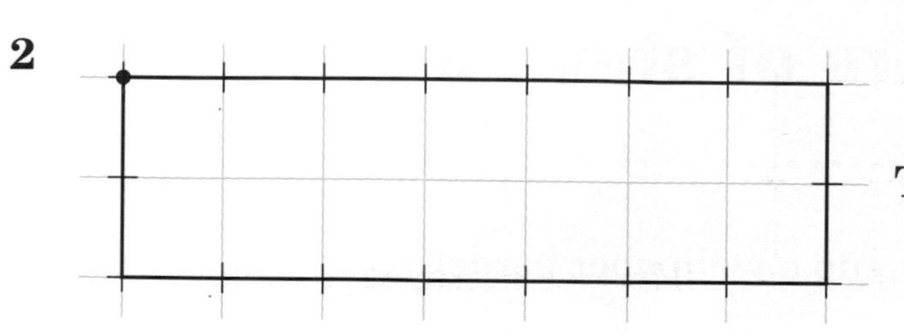

The perimeter is ...... cm

3

The perimeter is ...... cm

4

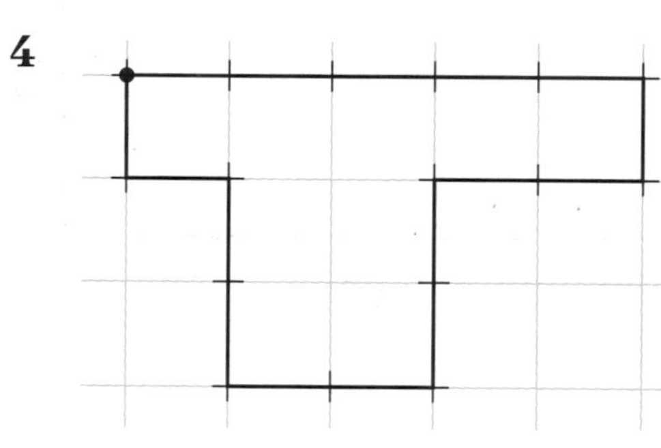

The perimeter is ...... cm

5

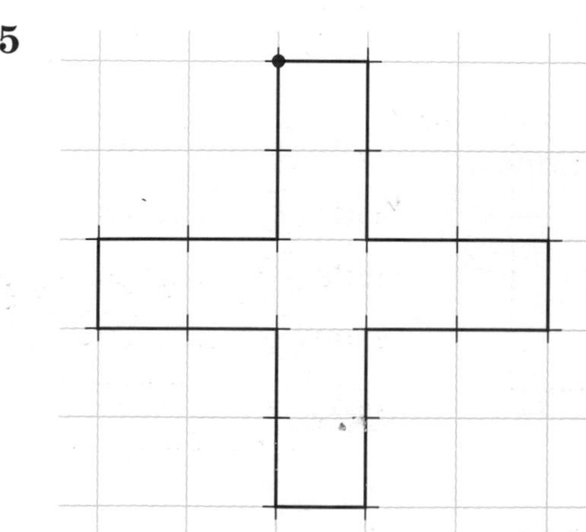

The perimeter is ...... cm

6

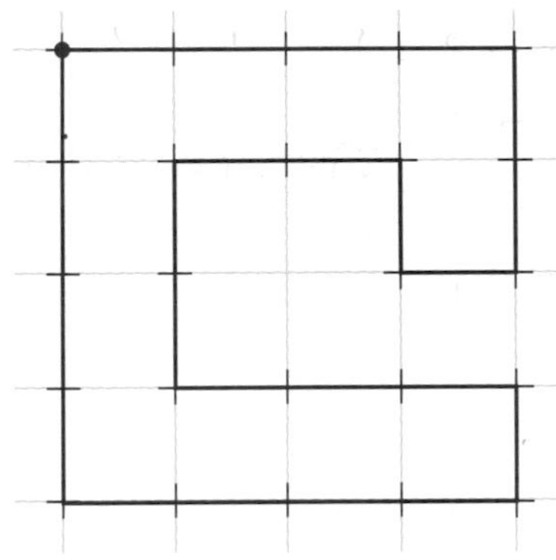

The perimeter is ...... cm

Example Find the perimeter of this shape by measuring.

You can find the perimeter by measuring each side and then adding them up.

side ①	6 cm
side ②	3 cm
side ③	6 cm
side ④	3 cm
Total	18 cm

①

④ ②

③

The perimeter of the shape is 18 cm.

H 5 **7** Find the perimeter of these shapes by measuring.

a

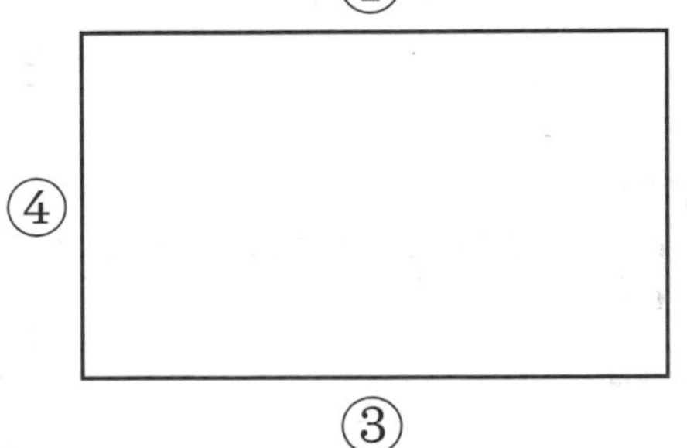

side ①	...... cm
side ②	...... cm
side ③	...... cm
side ④	...... cm
Total	

The perimeter of the shape is ...... cm

b

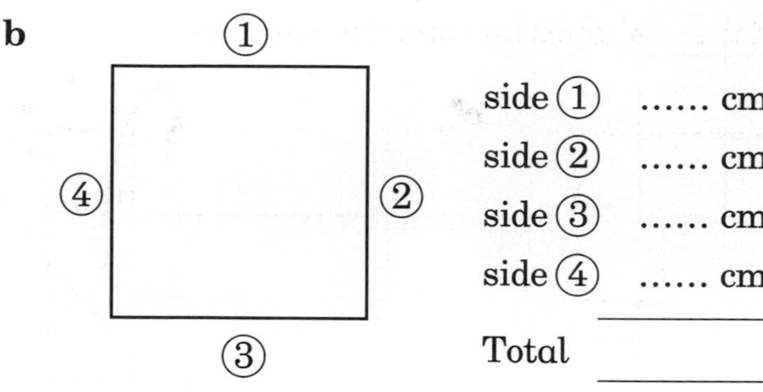

The perimeter of the shape is ...... cm

c

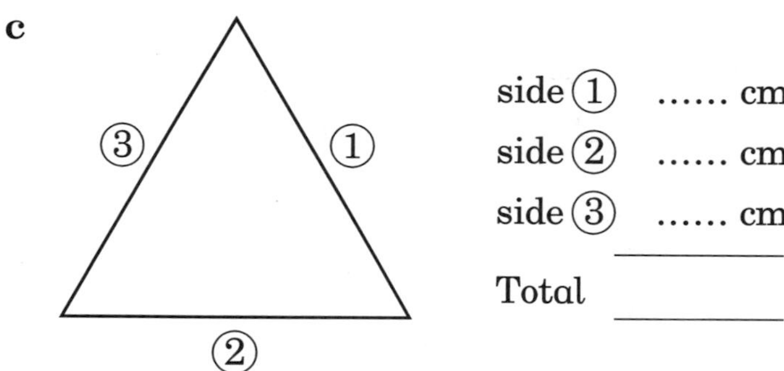

The perimeter of the shape is ...... cm

d

①
⑥ ②
⑤ ③
④

side ① ...... cm
side ② ...... cm
side ③ ...... cm
side ④ ...... cm
side ⑤ ...... cm
side ⑥ ...... cm

Total ______

1 Find the area of these shapes by counting squares.

a

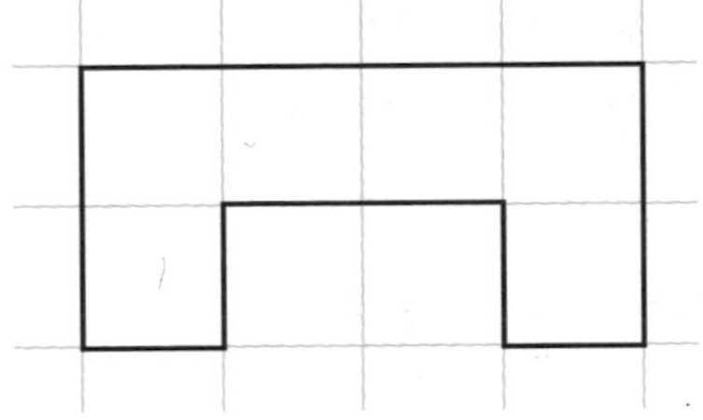

c

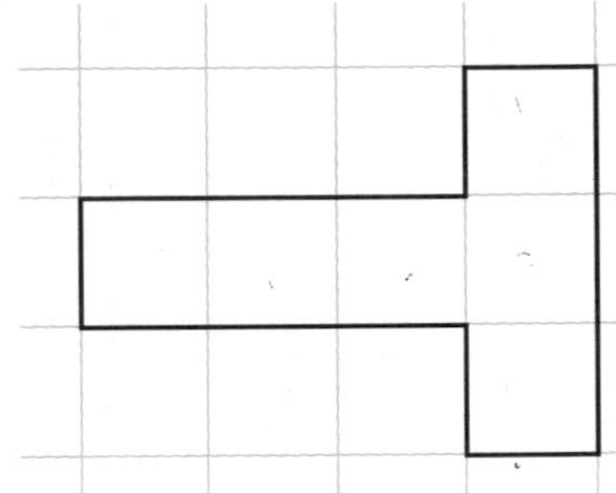

b

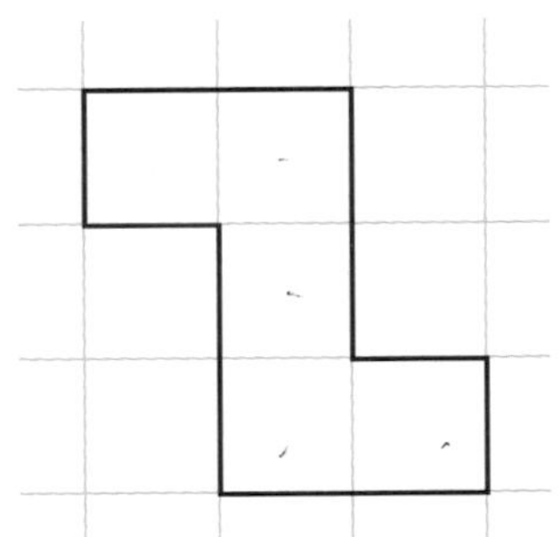

d

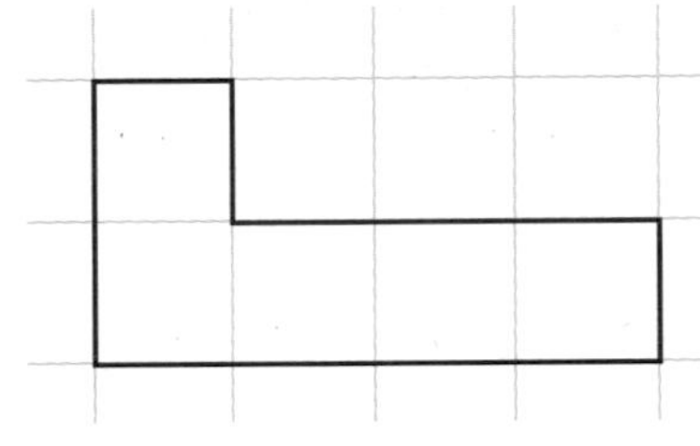

2 Find the area of these shapes.

a

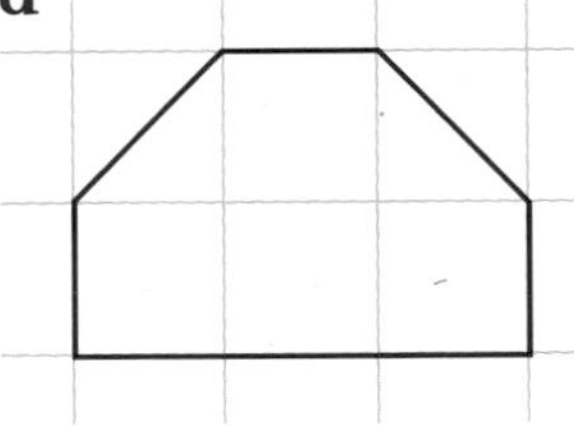

c

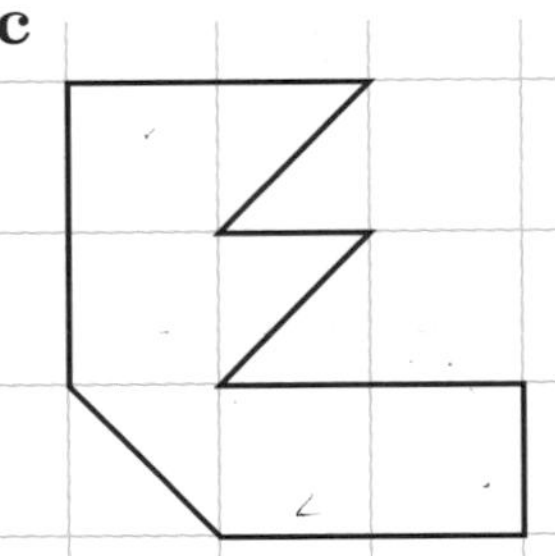

e

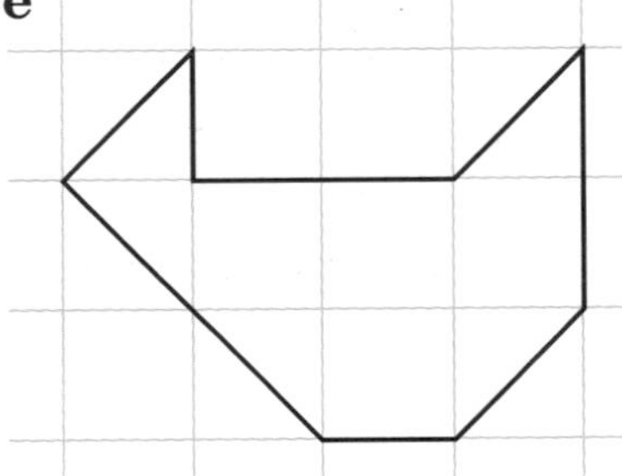

b

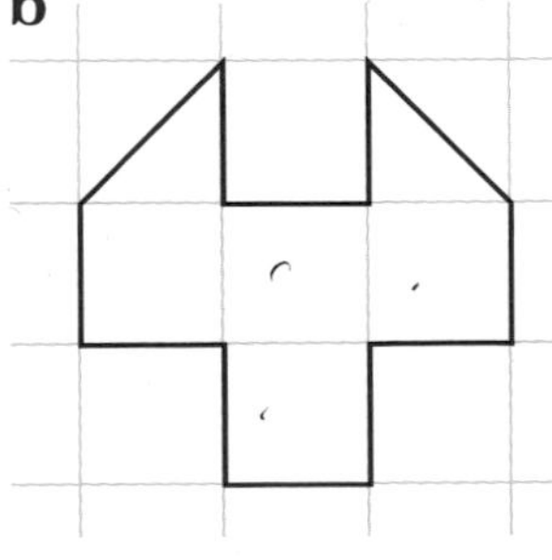

d

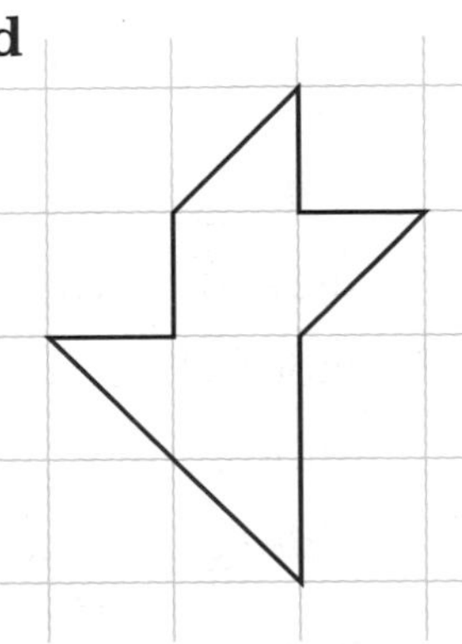

f

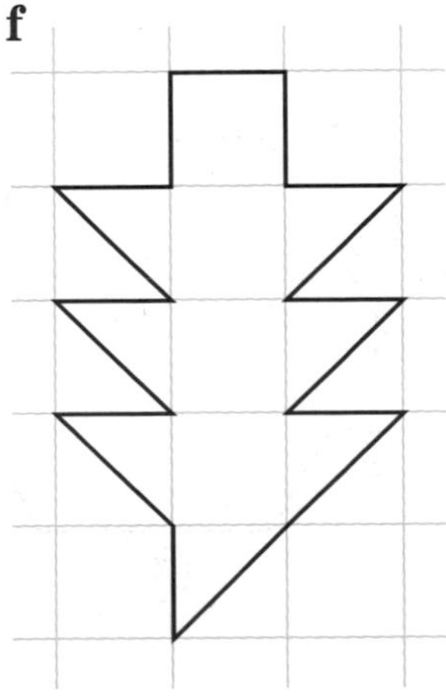

7 **3** Find the approximate area of these shapes.
Count the whole squares.
Colour the parts which are more than half a square.

a

Whole squares + parts more than half a square = approximate area

.... + ... = ...

The approximate area is .... cm^2

b

Whole squares + parts more than half a square = approximate area

.... + ... = ...

The approximate area is .... cm^2

4 Find the area of these rectangles by multiplying.

a

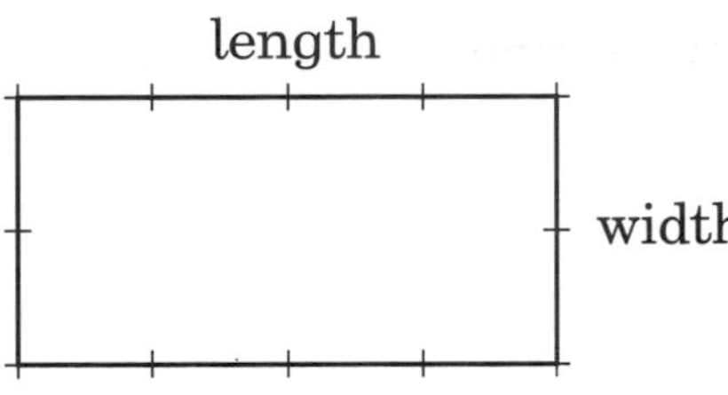

Length = ……

Width = ……

Area = …… × ……

Area = …… cm^2

b

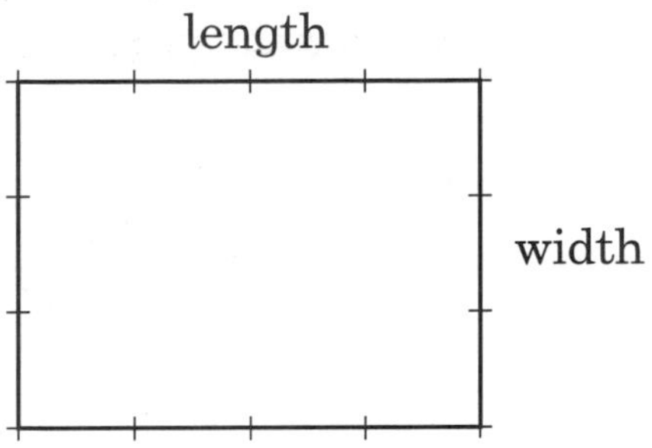

Length = ……

Width = ……

Area = …… × ……

Area = …… cm^2

5 Find the perimeter of these shapes.

a

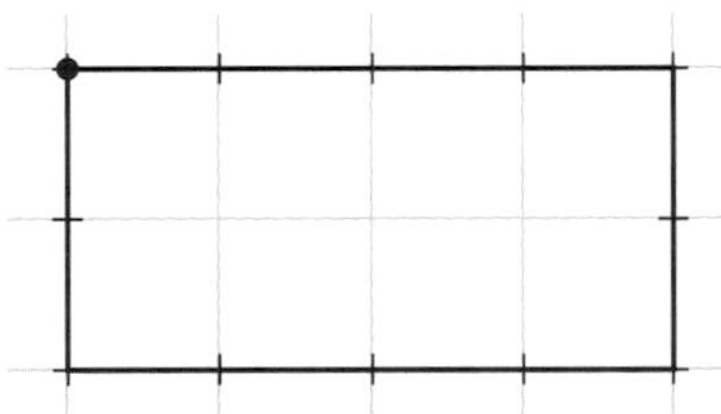

The perimeter is …… cm

b

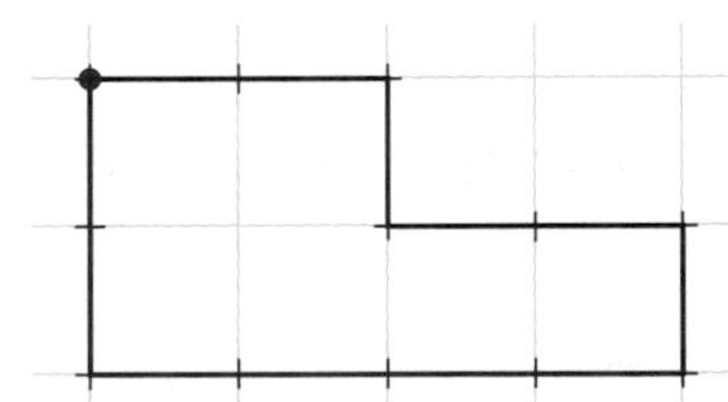

The perimeter is …… cm

6 Find the perimeter of this shape by measuring.

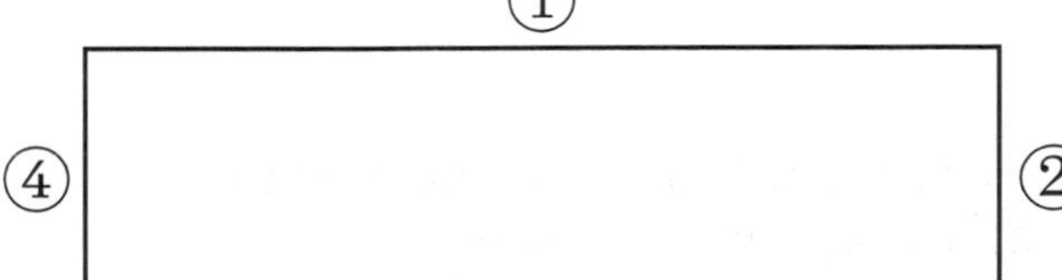

side ① …… cm

side ② …… cm

side ③ …… cm

side ④ …… cm

Total ______

The perimeter of the shape is …… cm

Division 3

Copy these into your book.
Work out the answers.

Exercise 1

1 $22 \div 2$

2 $30 \div 2$

3 $24 \div 2$

4 $28 \div 2$

5 $26 \div 2$

6 $42 \div 3$

7 $36 \div 3$

8 $45 \div 3$

9 $39 \div 3$

10 $33 \div 3$

Exercise 2

1 $2\overline{)42}$

2 $2\overline{)68}$

3 $2\overline{)64}$

4 $2\overline{)44}$

5 $2\overline{)86}$

6 $3\overline{)63}$

7 $3\overline{)93}$

8 $3\overline{)69}$

9 $3\overline{)96}$

10 $3\overline{)99}$

Exercise 3

1 $55 \div 5$

2 $75 \div 5$

3 $60 \div 5$

4 $70 \div 5$

5 $65 \div 5$

6 $110 \div 10$

7 $140 \div 10$

8 $130 \div 10$

9 $150 \div 10$

10 $120 \div 10$

Exercise 4

1 $5\overline{)80}$

2 $5\overline{)555}$

3 $5\overline{)95}$

4 $5\overline{)255}$

5 $5\overline{)155}$

6 $10\overline{)330}$

7 $10\overline{)180}$

8 $10\overline{)470}$

9 $10\overline{)210}$

10 $10\overline{)560}$

15 Fractions and percentages

In his will a man left his 17 camels to his three sons.

He left $\frac{1}{2}$ of the camels to his oldest son, $\frac{1}{3}$ of them to his middle son and $\frac{1}{9}$ of them to the youngest son.

When the boys came to divide up the camels they found that 17 was a very awkward number!

The youngest boy who was the cleverest had a good idea. He borrowed a camel from their neighbour so that they now had 18.

They now split up the camels: $\frac{1}{2}$ of $18 = 9$
$\frac{1}{3}$ of $18 = 6$
$\frac{1}{9}$ of $18 = 2$

$9 + 6 + 2 = 17$ camels so they could return the last camel to their neighbour!!

Can you find out how this trick works?

1 Introducing fractions

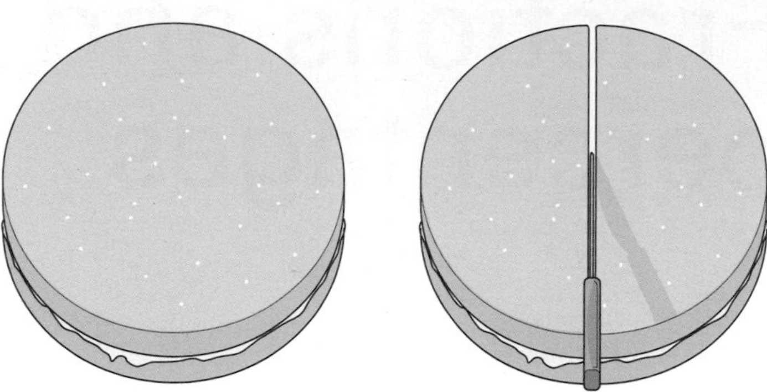

Whole This is a **whole**.

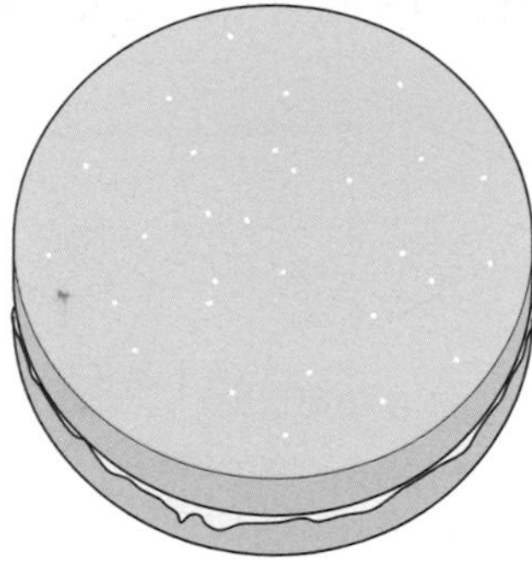

Half A **half** means dividing something into 2 equal pieces.

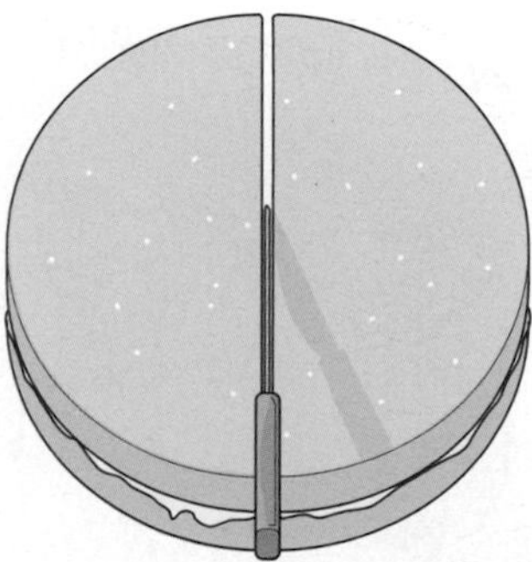

This cake is divided in **half**.

Exercise 15:1

1 Which of the following cakes have been cut in half?

a

c

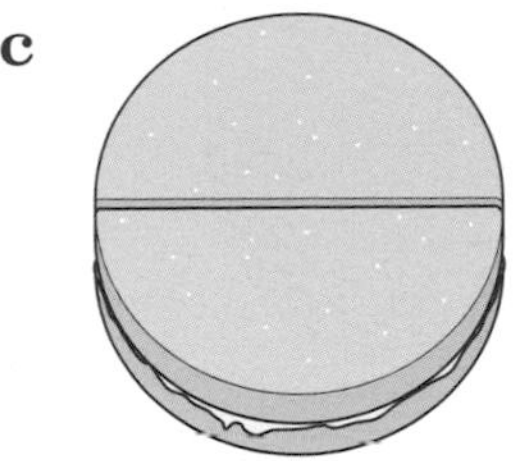

e

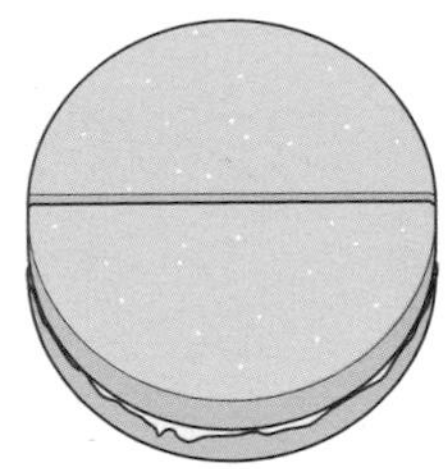

b

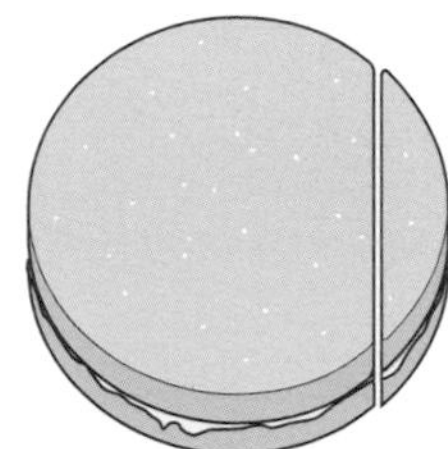

d

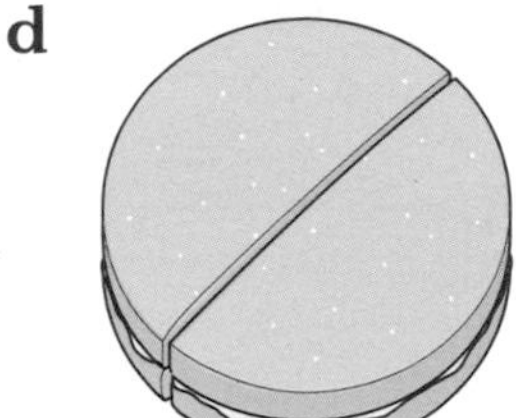

f

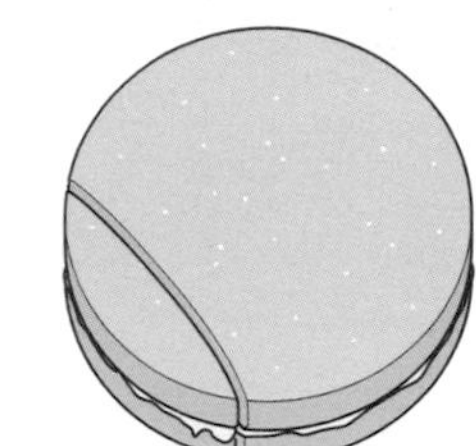

2 Which of the following sandwiches have been cut in half?

a

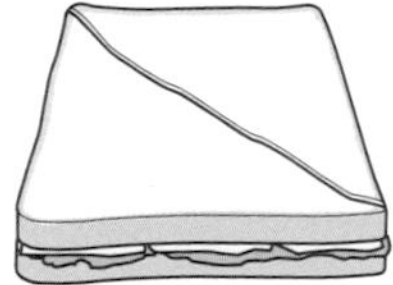

c

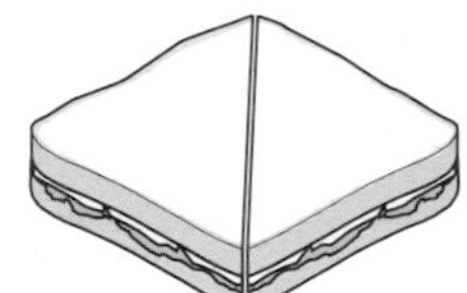

e

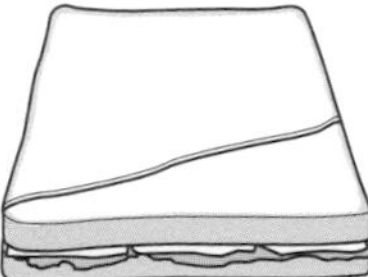

b

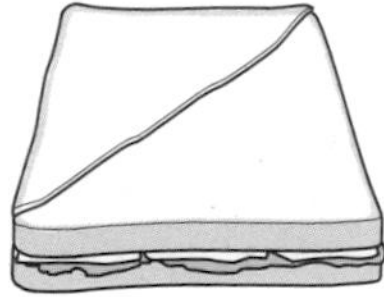

d

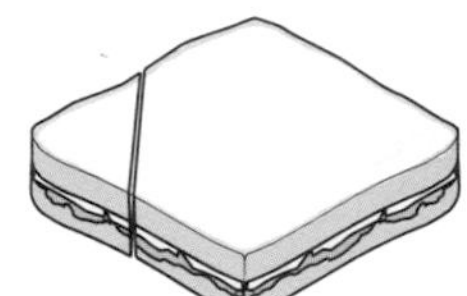

f

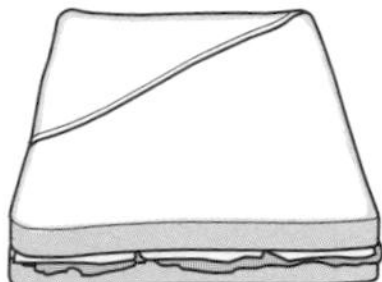

3 Which of the following pies have been cut in half?

a

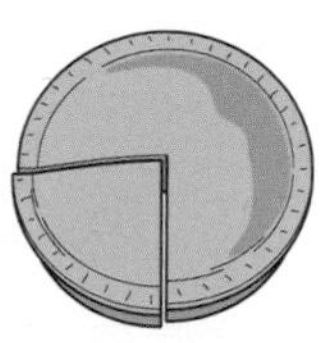

c

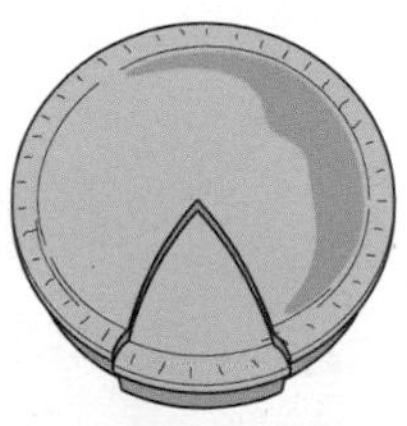

e

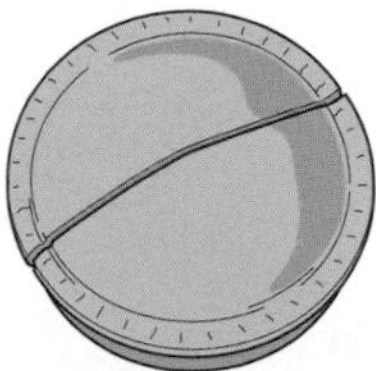

b

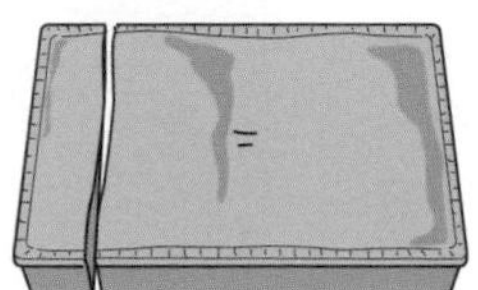

d

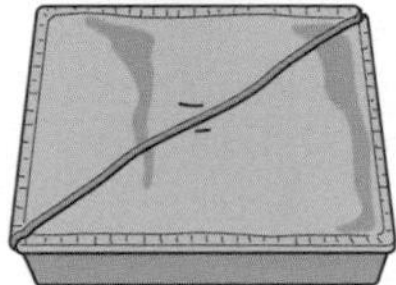

f

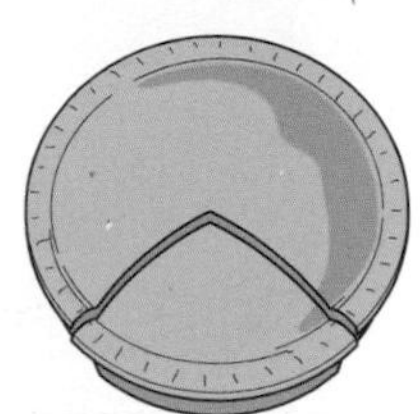

G 1

This is a whole.

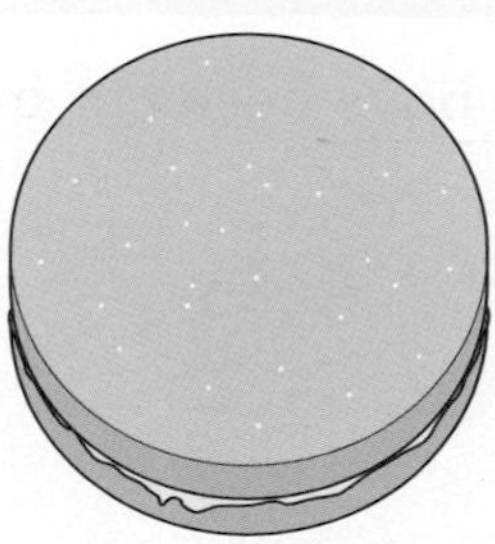

Quarter A **quarter** means dividing something into 4 equal pieces.

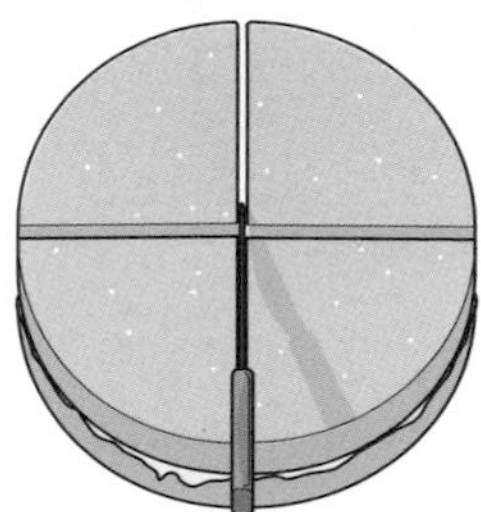

This cake is divided in **quarters**.

Exercise 15:2

1 Which of the following cakes have been cut into quarters?

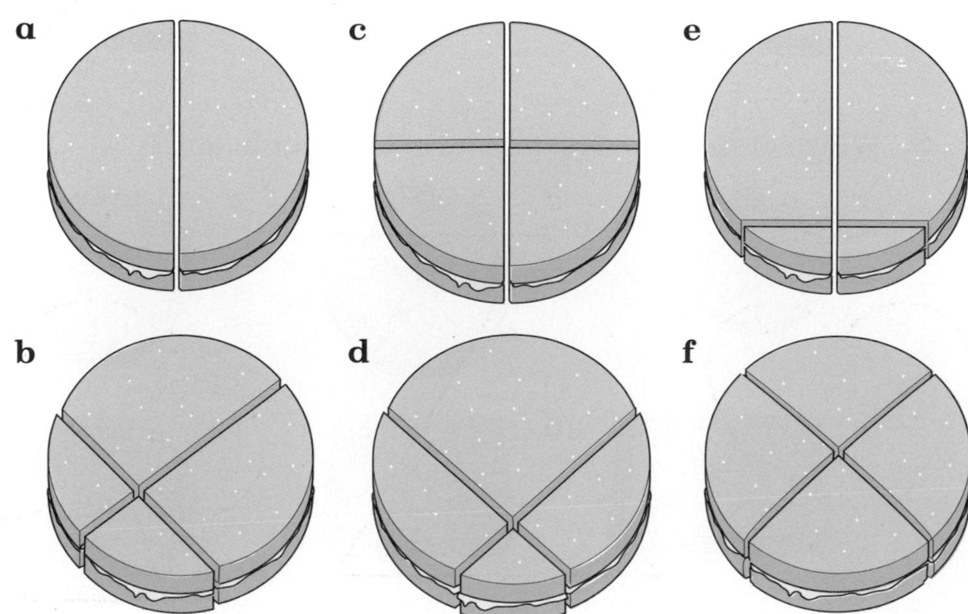

2 Which of the following sandwiches have been cut into quarters?

a

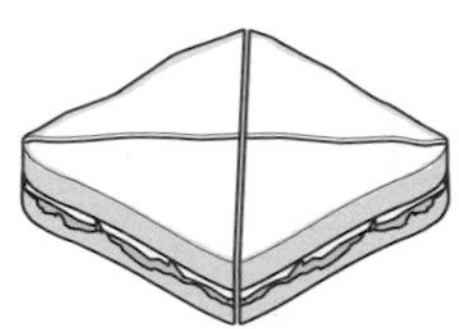

b

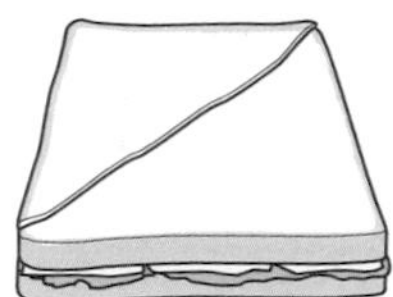

c

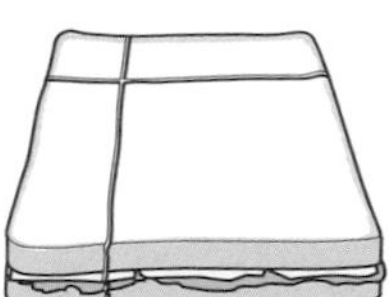

3 Which of the following pies have been cut into quarters?

a

b

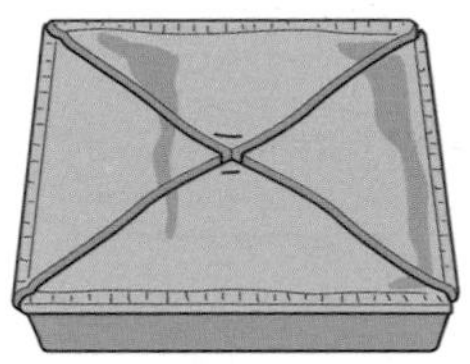

c

G 2

Exercise 15:3

W 1

1 You need copies of these shapes.
Shade in half of each of the following shapes.

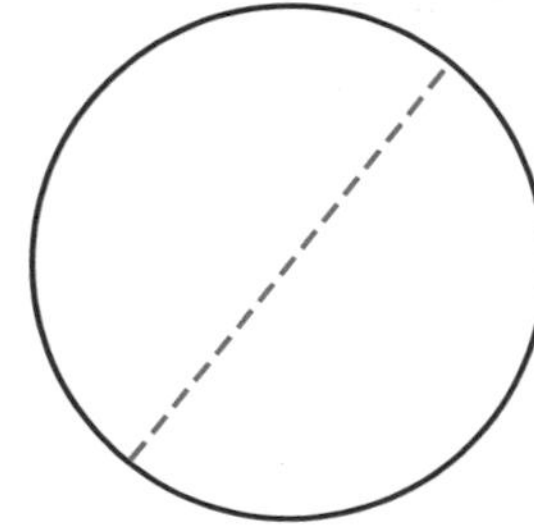

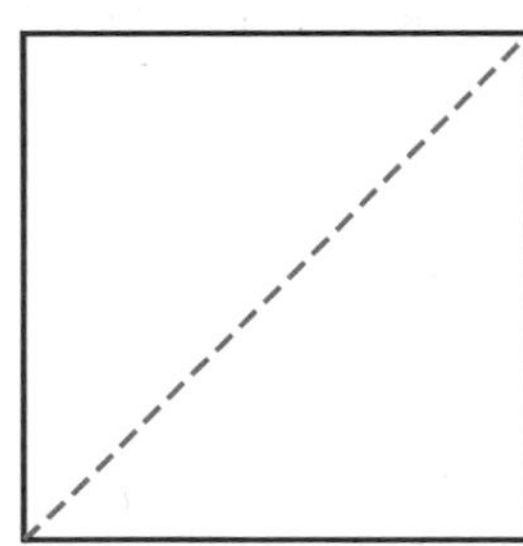

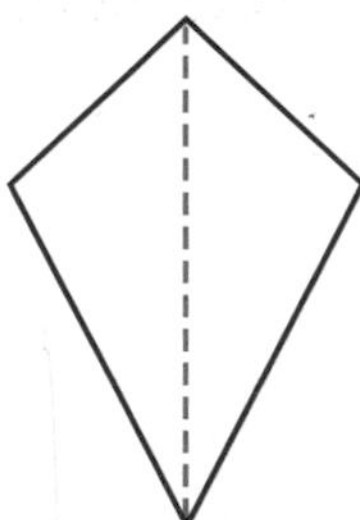

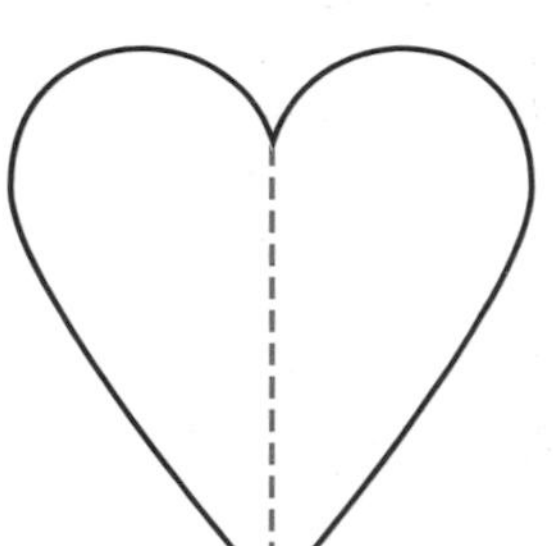

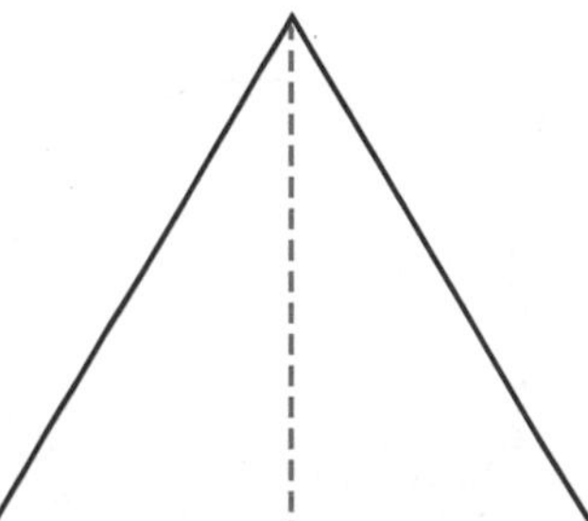

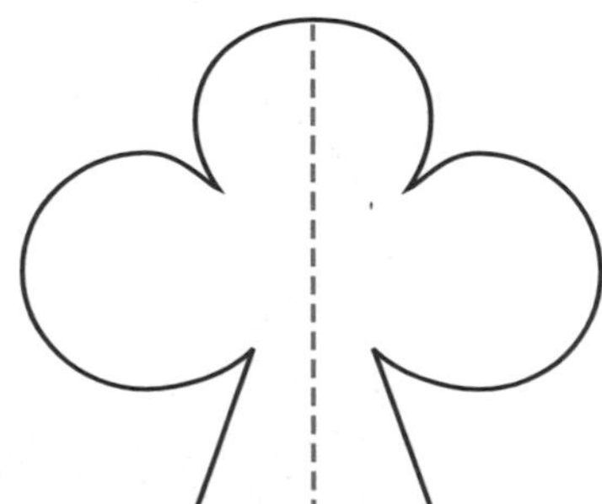

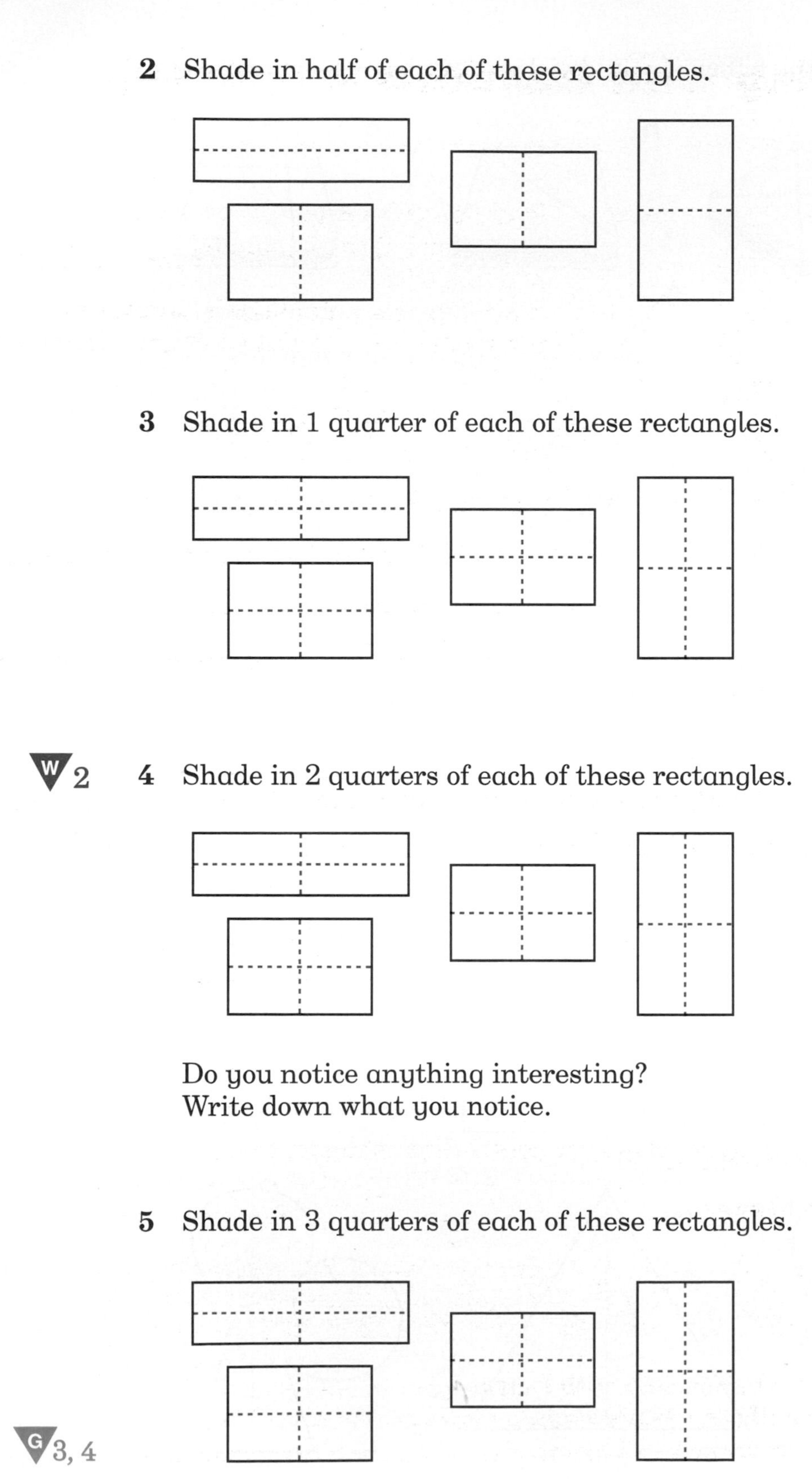

2 Shade in half of each of these rectangles.

3 Shade in 1 quarter of each of these rectangles.

W 2

4 Shade in 2 quarters of each of these rectangles.

Do you notice anything interesting?
Write down what you notice.

5 Shade in 3 quarters of each of these rectangles.

G 3, 4

2 Working with fractions

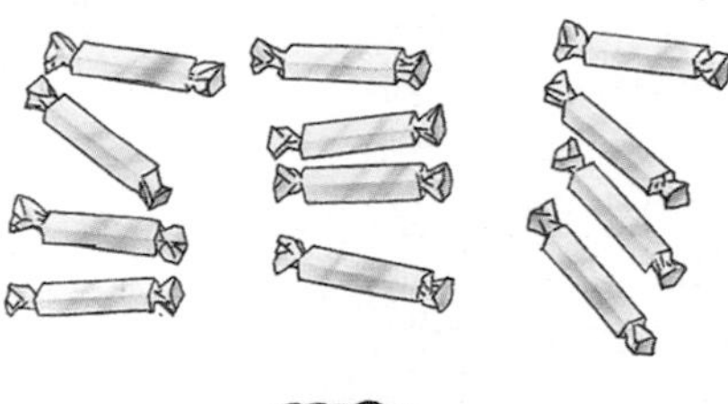

Paul has 12 sweets.
He is sharing them equally with his friend.
Each boy gets half of the sweets.

Finding a half

By sharing the sweets into 2 equal groups you are **finding a half**.

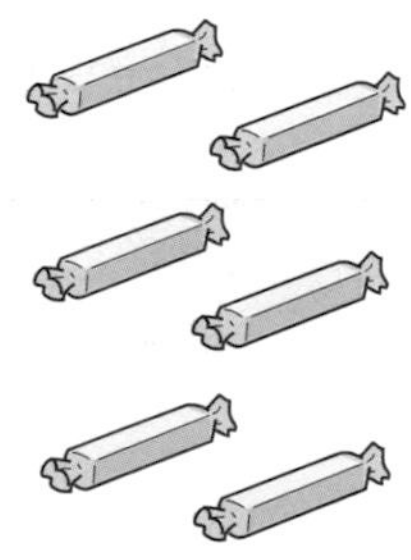
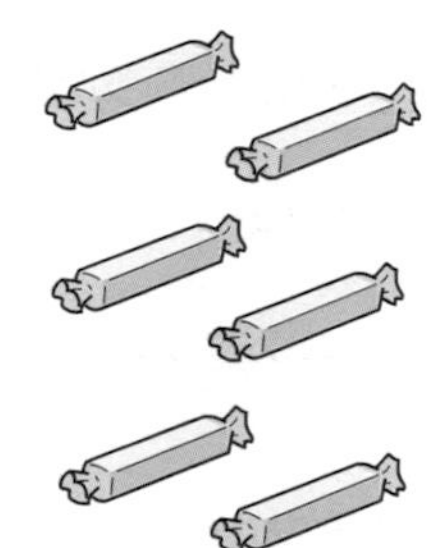

Exercise 15:4

You could use counters to help you.

1 Jenny has 10 sweets.

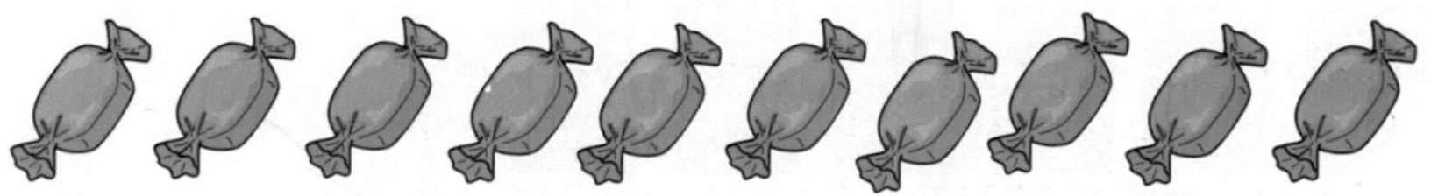

Jenny shares her sweets with Laura.
They have half the sweets each.
How many does each girl have?

2 Mitchell has 6 biscuits.

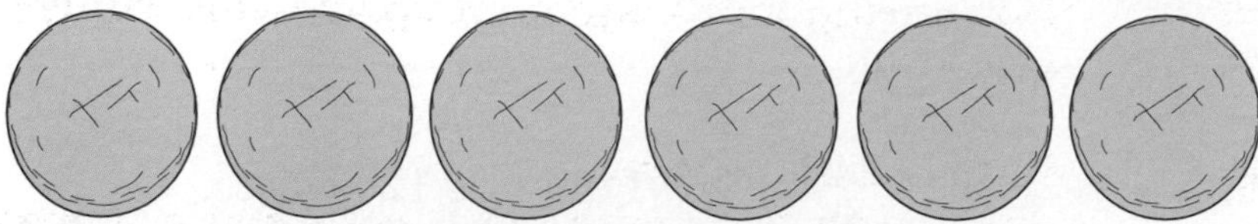

Mitchell shares his biscuits with Aaron.
They have half the biscuits each.
How many biscuits does each person have?

3 Mark has 4 cans of lemonade.

Mark shares his lemonade with Simon.
They have half the cans of lemonade each.
How many cans of lemonade does each boy have?

4 Diane has 8 bars of chocolate.

Diane shares her chocolate with Wendy.
They have half the bars of chocolate each.
How many bars of chocolate does each girl have?

5 Joe has 12 football cards.

Joe shares his football cards with Jack.
They have half the cards each.
How many cards does each boy have?

Finding a quarter

By sharing into 4 equal groups you are **finding a quarter**.

Example

Lucy has 16 sweets.
She shares them with her friends Katie, James and Joanne.
They get a quarter of the sweets each.

Exercise 15:5

You could use counters to help you.

1 Jessica has 12 biscuits.

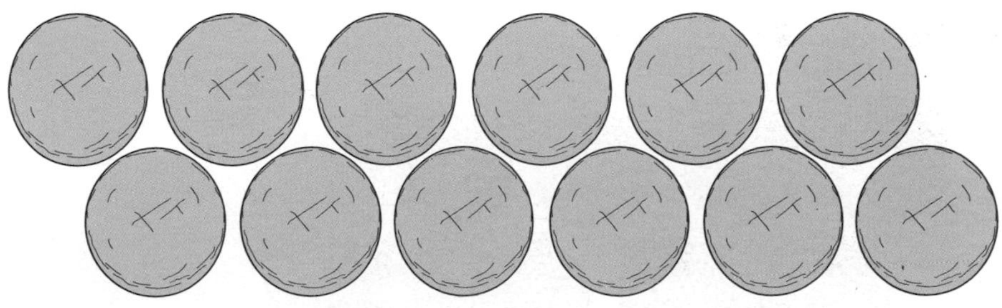

She shares them with her 3 friends.
They have a quarter of the biscuits each.
How many biscuits does each person have?

2 Chris has 8 bars of chocolate.

He shares them with his three friends.
They have a quarter of the bars of chocolate each.
How many bars of chocolate does each person have?

3 Pauline has 4 sweets.

She shares her sweets with three friends.
They have a quarter of the sweets each.
How many sweets does each person have?

4 Daniel has 20 football cards.

He shares them with his three friends.
They have a quarter of the football cards each.
How many football cards does each person have?

5 Sarah has 8 doughnuts.

She shares them with her three friends.
They have a quarter of the doughnuts each.
How many doughnuts does each child have?

Example Louise has 12 cakes.

She shares them with her three friends.
They each have a quarter of the cakes.
How many cakes does each person have?
How many cakes did Louise give away?

9 is three quarters of 12.
Louise divided her 12 cakes into quarters.
She kept one quarter.

She gave away three quarters.

$\frac{3}{4}$ of 12 is 9

6 Rhiannon has 8 sweets.

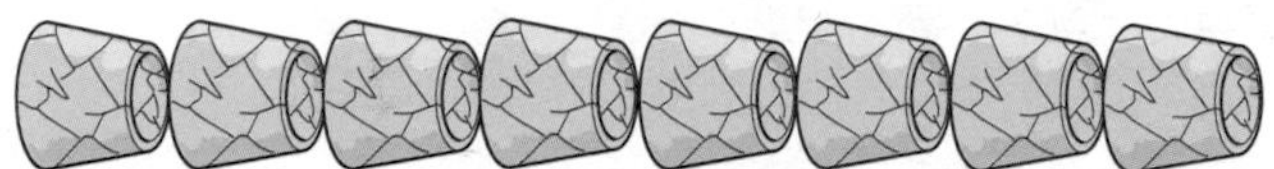

She shares them with her three friends.
They each have one quarter of the sweets.
How many sweets does Rhiannon have?
How many sweets does she give away?

3 Introducing percentages

The Roman army had soldiers called centurions.
They marched in a square of 10 soldiers by 10 soldiers.

Percentage

Percentage means out of 100.
You write it like this %
13% means thirteen out of one hundred.

Exercise 15:6

1 How many of these centurions are wearing red helmets?

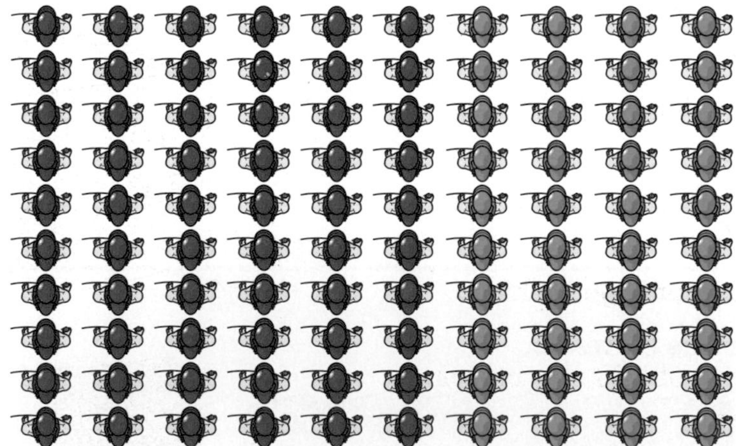

a Write this in words out of one hundred.
Write it as a percentage %

b How many of the centurions are wearing blue helmets?
Write it as a percentage.

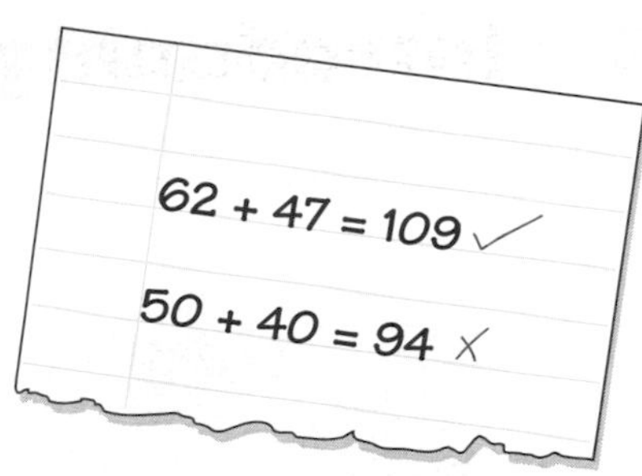

2 Mary did a maths test.
There were 100 questions.
She got 63 questions correct.

a What percentage is this?

b How many did Mary get wrong?

c What percentage is this?

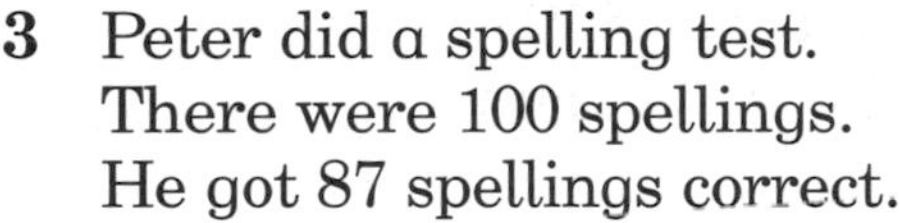

3 Peter did a spelling test.
There were 100 spellings.
He got 87 spellings correct.

a What percentage is this?

b How many did Peter get wrong?

c What percentage is this?

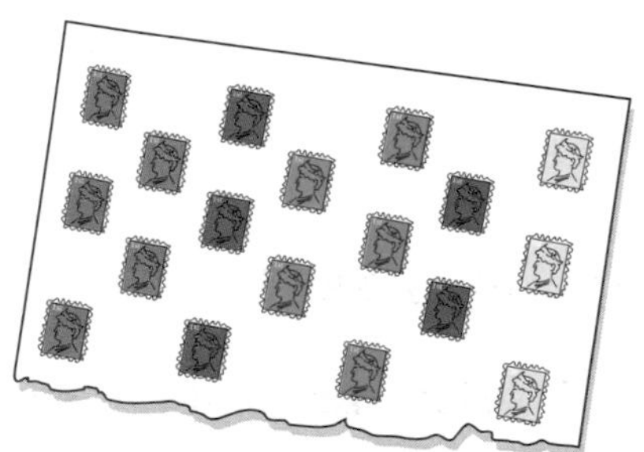

4 Sarah has 100 stamps.
28 are from England.

a What percentage is this?

b How many stamps are not from England?

c What percentage is this?

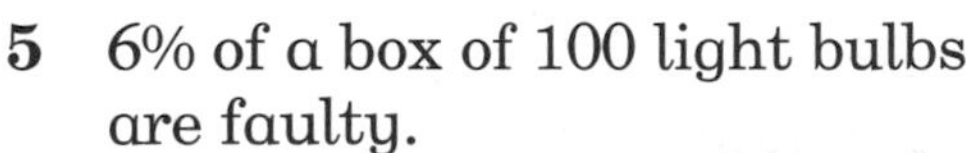

5 6% of a box of 100 light bulbs are faulty.

a How many light bulbs are faulty?

b How many light bulbs are not faulty?

c What percentage is this?

6 36% of a box of 100 chocolates were toffees.

a How many chocolates were toffees?

b How many chocolates were not toffees?

5, 6

c What percentage is this?

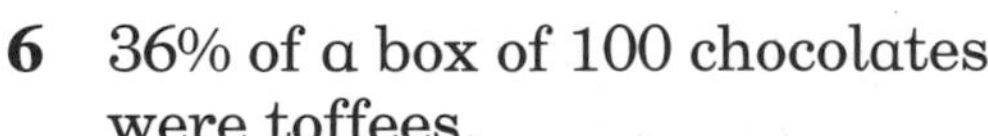

4 Fractional parts

Whose offer is better?

Exercise 15:7

W 3 Here are some rectangles.

1 **a** Colour in half the squares on this rectangle.

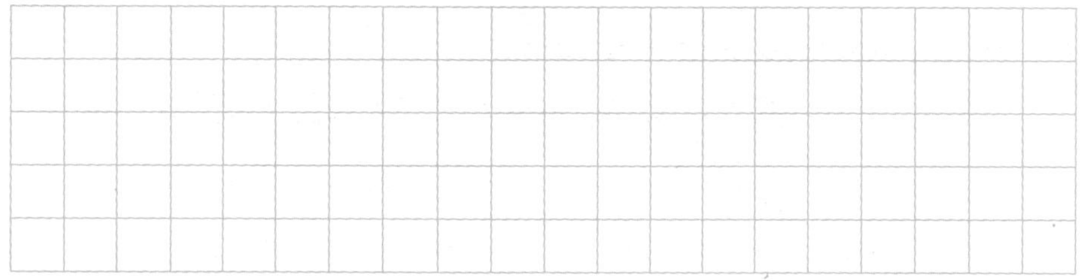

b How many squares are there?

c How many squares have you coloured in?

d Write this as a percentage.

$\frac{1}{2}$ is equal to ......%

W 3 **2** **a** Colour in one quarter of the squares on this rectangle.

b How many squares are there?

c How many squares have you coloured in?

d What percentage is this?

$\frac{1}{4}$ is equal to ......%

 3

3 **a** Colour in three quarters of the squares on this rectangle.

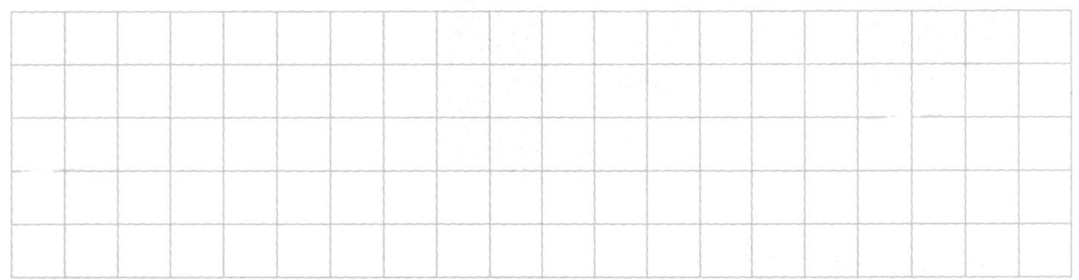

b How many squares are there?

c How many squares have you coloured in?

d Write this as a percentage.

$\frac{3}{4}$ is equal to ......%

4 Janet had 8 cans of lemonade.

a She gave 50% of the cans to her friend Claire.
How many cans did Claire get?

b How many cans did Janet have left?

5 Mansoor has 12 football cards.

a He gave 25% to his friend David.
How many football cards did David get?

G 7, 8 **b** How many football cards did Mansoor have left?

1 Which of the following cakes have been cut into half?

a

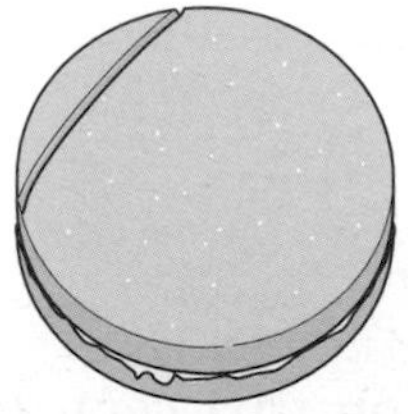

b

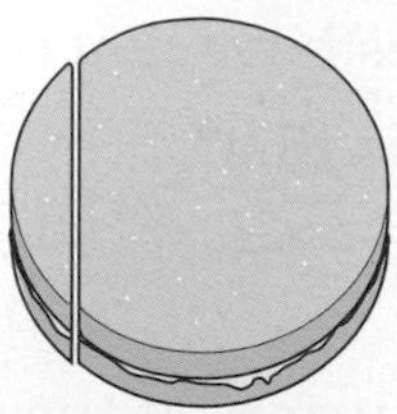

c

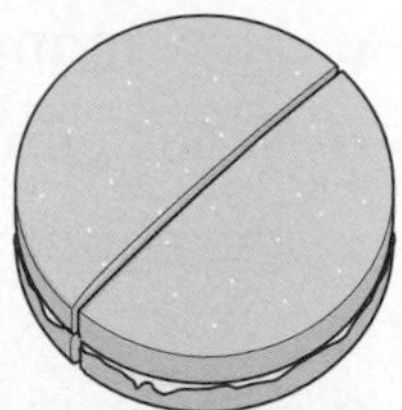

2 Which of the following cakes have been cut into quarters?

a

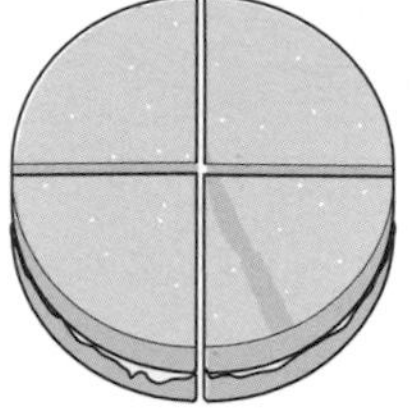

b

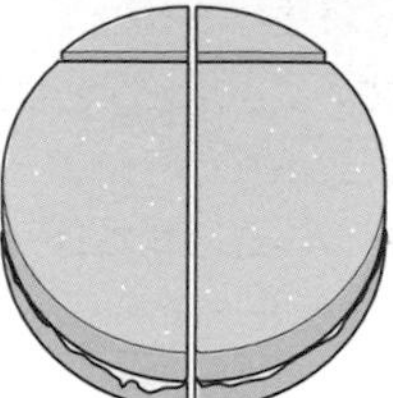

c

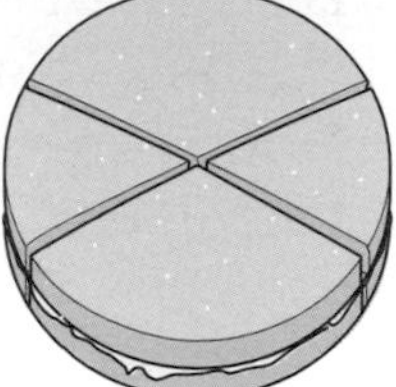

3 Mary has 10 sweets.

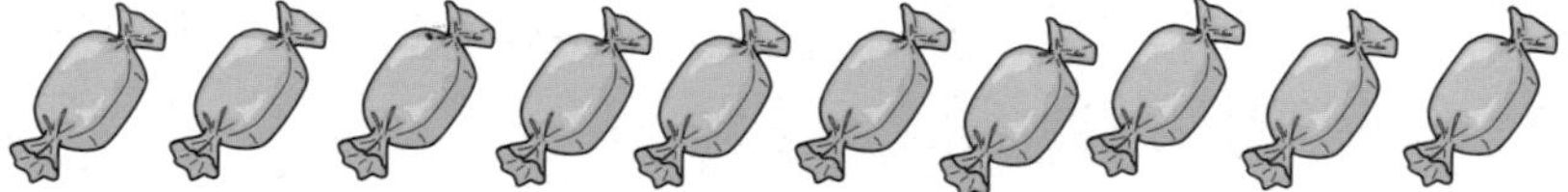

Mary gives half her sweets to her friend.
How many sweets does each girl have?

4 Jon has 6 cans of cola.

Jon gives half his cans of cola to his friend.
How many cans of cola does each boy have?

5 Caroline has 8 pencils.

She shares them equally with her friends.
They have a quarter of the pencils each.
How many pencils does each person have?

6 Lewis has 12 doughnuts.

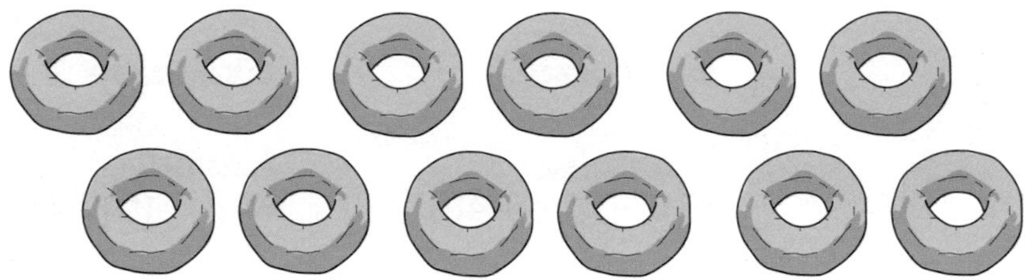

He shared them with 3 friends.
They have one quarter of the doughnuts each.
How many did Lewis keep?
How many did Lewis give away?

7 Here are 100 people.

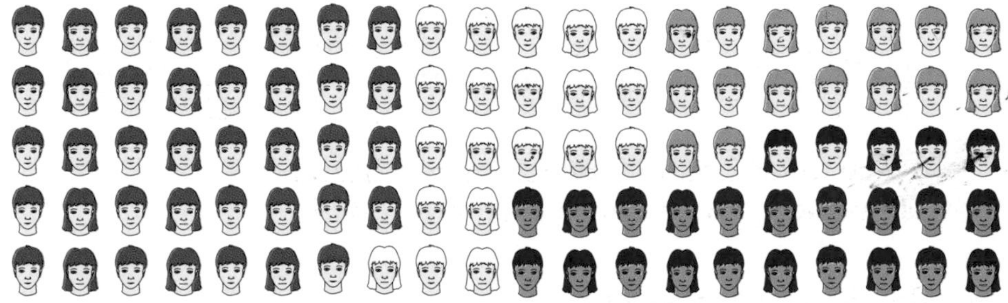

a How many of the people have black hair?
Write this as a percentage.

b How many of the people have red hair?
Write this as a percentage.

c How many people have blonde hair?
Write this as a percentage.

8 James has 100 stamps.
32 stamps are from France.

a What percentage is this?

b How many stamps are not from France?

c What percentage is this?

Money 3

Exercise 1

Example
70 p − 20 p = 50 p

Find the answer to each of these questions.

1 50 p − 30 p

2 60 p − 40 p

3 70 p − 30 p

4 90 p − 10 p

5 £1.20 − 30 p

6 £1.40 − 60 p

7 £1.50 − 80 p

8 £1.80 − 90 p

9 £2.30 − 50 p

10 £2.40 − 70 p

Exercise 2

Example
85 p − 45 p = 40 p

Find the answer to each of these questions.

1 55 p − 25 p

2 35 p − 15 p

3 45 p − 25 p

4 75 p − 35 p

5 85 p − 45 p

6 95 p − 15 p

7 95 p − 85 p

8 £1.25 − 15 p

9 £1.45 − 35 p

10 £1.85 − 55 p

Exercise 3

Example
60 p − 12 p = 48 p

Find the answer to each of these questions.

1 20 p − 8 p

2 40 p − 12 p

3 70 p − 24 p

4 90 p − 76 p

5 90 p − 19 p

6 £1.20 − 12 p

7 £1.30 − 28 p

8 £1.60 − 32 p

9 £1.70 − 57 p

10 £1.90 − 36 p

16 Statistics: what does it all mean?

CORE

1 **The mean**

2 **Other types of average**

3 **The range**

4 **Misleading statistics**

QUESTIONS

HELP YOURSELF **Time 3**

An average puzzle to start this chapter:

A stick is broken into two pieces at random.
What is the average length of the shorter piece?

1 The mean

Football clubs want to know how well they are performing. They want to know what their average goal score is.

Mean

The **mean** is a type of average.
To find the mean add up all the scores.
Then divide it by the number of scores.

Example

In their first three matches, Chester scored 2 goals, 3 goals and 2 goals.
Find the mean number of goals scored per match.

Number of goals $2 + 3 + 2 = 7$
Number of games $1 + 1 + 1 = 3$
Mean number of goals $7 \div 3 = 2.33...$
The mean number of goals scored is 2.33... per match.

Exercise 16:1

1 Find the mean number of goals scored by each of these teams in the first matches of the season.

a Arsenal 5, 0, 2, 0, 4
Number of goals $5 + 0 + 2 + 0 + 4 = ?$
Number of games $1 + 1 + 1 + 1 + 1 = ?$
Mean number of goals $? \div 5 = ?$

b Aston Villa 0, 2, 1, 2
Number of goals $? + ? + ? + ? = ?$
Number of games $1 + 1 + 1 + 1 = ?$
Mean number of goals $? \div 4 = ?$

c Barnsley 0, 2, 1, 0, 0

d Crystal Palace 0, 1, 0, 2

e Manchester United 2, 3, 2, 1, 0

f West Ham 2, 0, 3, 2

g Leeds 1, 0, 0, 0, 1

G 1, 2 h Newcastle 1, 1, 2, 1, 1

W 1 **2** In Question **1**, who has the highest mean goal score?

3 Here are 5 children from the pupil cards on Worksheet 1.

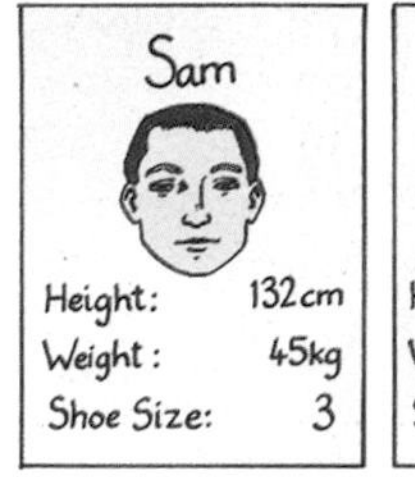

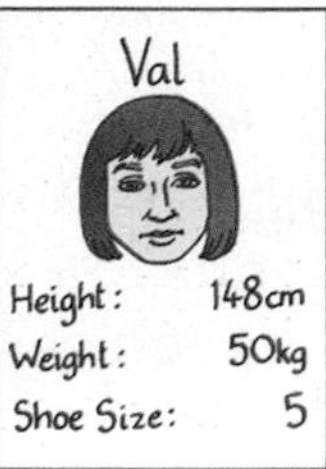

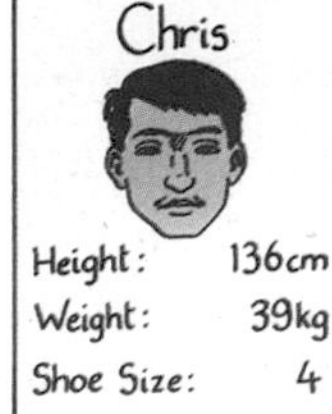

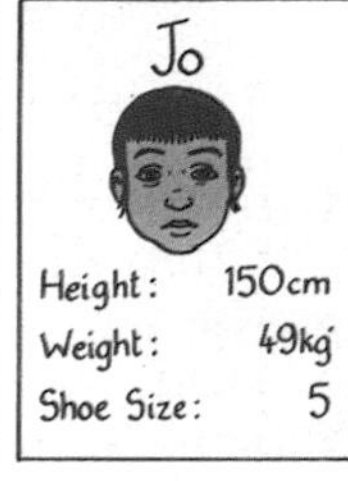

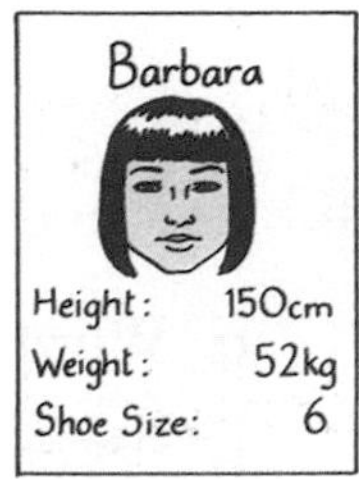

a What is the mean height of these 5 children?

b Which children are taller than the mean height?

c What is the mean weight of these 5 children?

d Which of the children are lighter than the mean weight?

e What is the mean shoe size of these 5 children?

f Can you buy shoes the size of the mean shoe size?

W 1 **4** You need the pupil cards on Worksheet 1.

a What is the mean height of the 13 children?

b Which children are taller than the mean height?

c What is the mean weight of the 13 children?

d Which children are lighter than the mean weight?

e What is the mean shoe size of the 13 children?

W2 **5** Here are 5 babies from the baby cards on Worksheet 2.

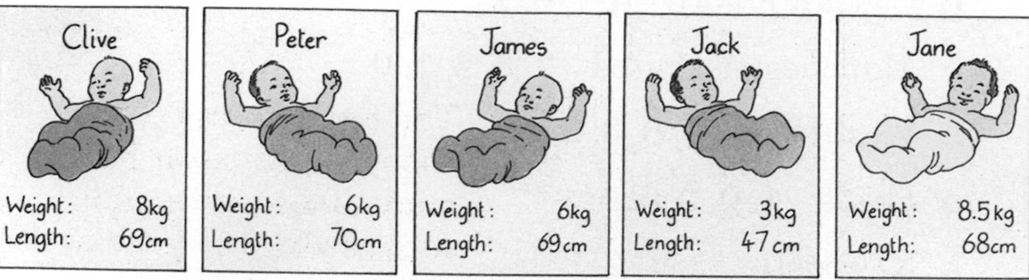

a What is the mean weight of these 5 babies?

b Which babies are heavier than the mean weight?

c What is the mean length of these 5 babies?

d Which babies are longer than the mean length?

W2 **6** You need the baby cards on Worksheet 2.

a What is the mean weight of the 13 babies?

b Which babies are heavier than the mean weight?

c What is the mean length of the 13 babies?

d Which babies are longer than the mean length?

7 David asked his friends how much money they had in their pockets.
He wrote down their answers.

99 p, 50 p, 25 p, 12 p, 52 p, 32 p, 90 p, 80 p

What is the mean amount of money?

8 Sarah asked her friends how much pocket money they received.
She wrote down their answers.

£1, £3, £2, £5, £2, £10, £8, £7

What is the mean amount of pocket money?

2 Other types of average

The manageress of this shoe shop looks at how many shoes are sold in each size.
She writes down the size that sells the most.
For women, this is size 5.
For men, it is size 9.

Mode The **mode** is the most common or the most popular value.
This is sometimes called the modal value.

Example The manageress of the shoe shop sells shoes in the following sizes:

5, 6, 4, 3, 5, 7, 5

The size that sells the most is size 5.
The mode is size 5.

Exercise 16:2

W1 **1** Here are 5 children from the pupil cards on Worksheet 1.

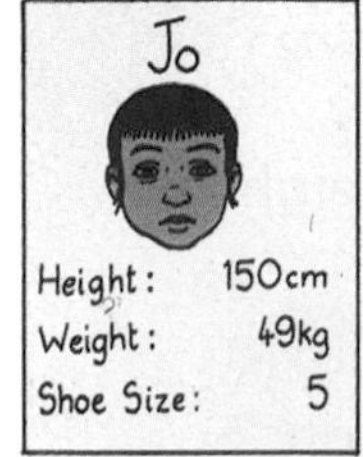

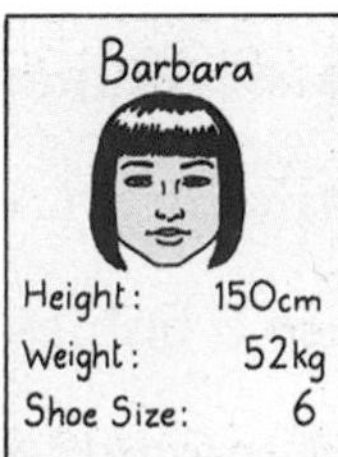

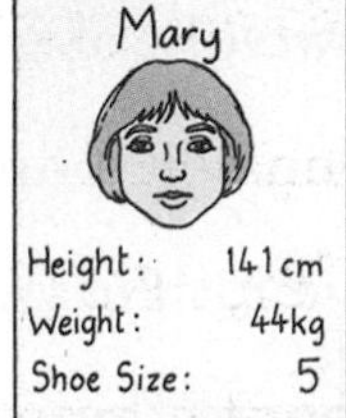

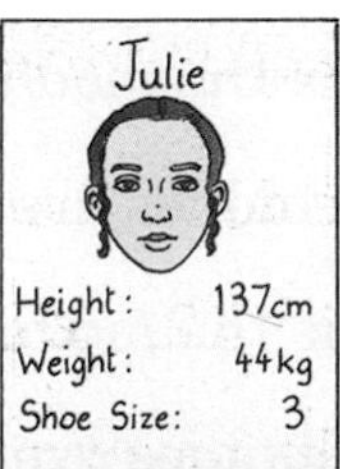

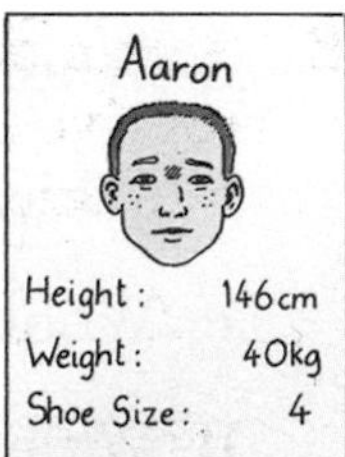

a What is the modal shoe size of these 5 children?

b Which children have shoes that are bigger than the modal shoe size?

c What is the modal weight of these 5 children?

d Which children are heavier than the modal weight?

e What is the modal height of these 5 children?

f Which children are taller than the modal height?

W1 **2** You need the pupil cards on Worksheet 1.

a What is the modal height of these 13 children?

b Which children are taller than the modal height?

c What is the modal weight of these 13 children?

d Which children are heavier than the modal weight?

e What is the modal shoe size of these 13 children?

f Which children have shoes that are bigger than the modal shoe size?

W2 **3** Here are 5 babies from the baby cards on Worksheet 2.

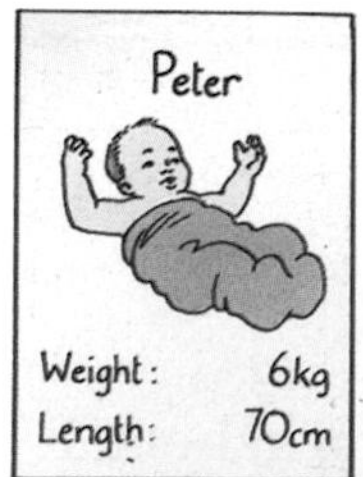

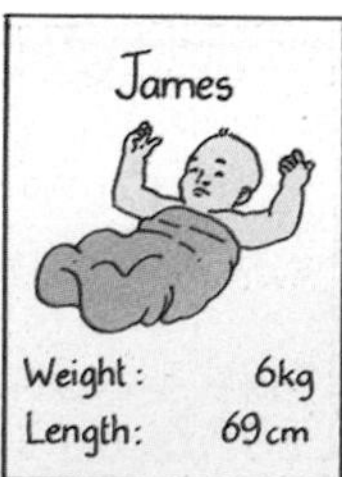

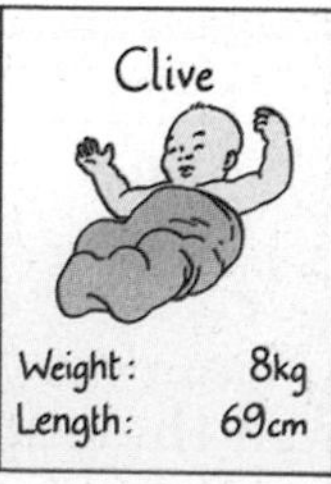

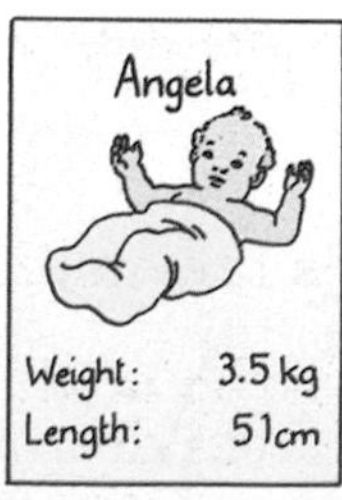

a What is the modal weight of these 5 babies?

b Which babies are heavier than the modal weight?

c What is the modal length of these 5 babies?

d Which babies are longer than the modal length?

 2 **4** You need the baby cards on Worksheet 2.

a What is the modal weight of these 13 babies?

b What is the modal length of these 13 babies?

Median Another type of average is called the **median**.
The median is the middle value of the data when it is arranged in order, smallest number first.

Example Liverpool scored 3, 1 and 2 goals in their first three matches of a season.
Find the median number of goals scored.

Arrange the number of goals in order 1, 2, 3
The median is 2, the number in the middle.
The median number of goals scored is 2.

Exercise 16:3

W 1 **1** Here are 5 children from the pupil cards on Worksheet 1.

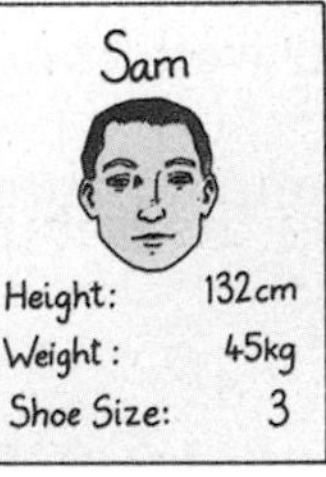

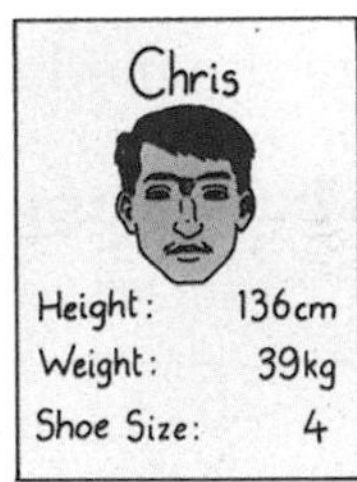

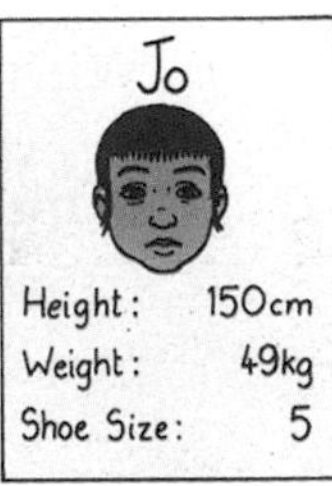

a What is the median height of these 5 children?

b What is the median weight of these 5 children?

c What is the median shoe size of these 5 children?

W 1 **2** You need the pupil cards on Worksheet 1.

a What is the median height of these 13 children?

b What is the median weight of these 13 children?

c What is the median shoe size of these 13 children?

W2 **3** Here are 5 babies from the baby cards on Worksheet 2.

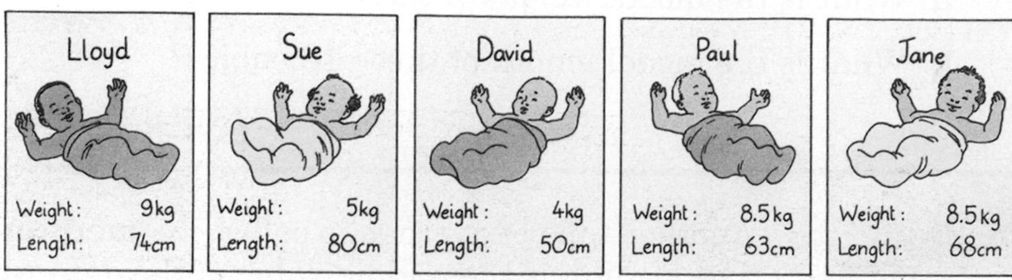

a What is the median length of these 5 babies?

b Which babies are shorter than the median length?

c What is the median weight of these 5 babies?

d Which babies are heavier than the median weight?

W2 **4** You need the baby cards on Worksheet 2.

a What is the median length of these 13 babies?

b Which babies are shorter than the median length?

c What is the median weight of these 13 babies?

d Which babies are heavier than the median weight?

5 Manchester United scored 2, 3, 2, 1, 0 goals in five matches.

a Arrange these in order, smallest first.

b What is the median number of goals scored?

6 Leeds scored 1, 0, 0, 1, 0 goals in five matches.

a Arrange these in order, smallest first.

b What is the median number of goals scored?

3 The range

Angus
Weight: 7 kg

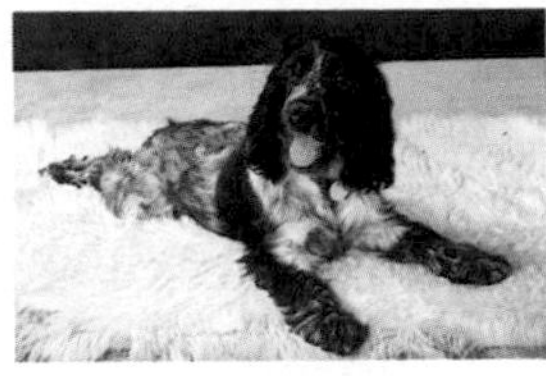

Barney
Weight: 6 kg

Crossley
Weight: 18 kg

Range

Crossley is the heaviest. Barney is the lightest.
The difference in their weights is

$18 - 6 = 12\,\text{kg}$

This is called the **range** of their weights.
The range is 12 kg.

Exercise 16:4

W 1 **1** Here are 4 children from the pupil cards on Worksheet 1.

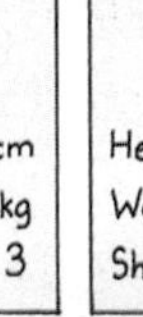

a Who is the tallest pupil?

b Who is the smallest pupil?

c What is the range of their heights?

d Who is the heaviest pupil?

e Who is the lightest pupil?

f What is the range of their weights?

g Who has the largest feet?

h Who has the smallest feet?

i What is the range of their shoe sizes?

W 1 **2** You need the pupil cards on Worksheet 1.

a Who is the tallest pupil?

b Who is the smallest pupil?

c What is the range of their heights?

d Who is the heaviest pupil?

e Who is the lightest pupil?

f What is the range of their weights?

g Who has the largest feet?

h Who has the smallest feet?

i What is the range of their shoe sizes?

W 2 **3** Here are 3 babies from the baby cards on Worksheet 2.

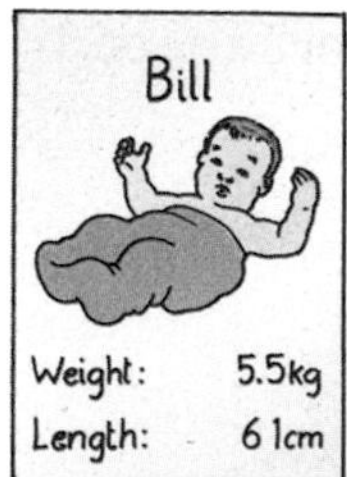

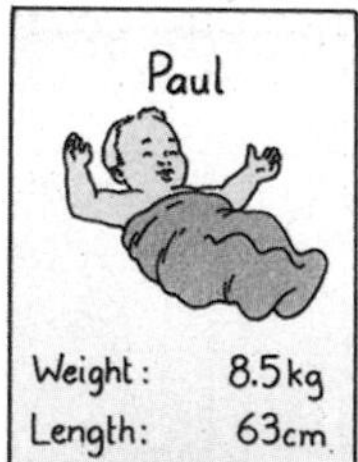

a Which is the longest baby?

b Which is the shortest baby?

c What is the range of their lengths?

d Which is the heaviest baby?

e Which is the lightest baby?

f What is the range of their weights?

W 2 **4** You need the baby cards on Worksheet 2.

a Which is the longest baby?

b Which is the shortest baby?

c What is the range of their lengths?

d Which is the heaviest baby?

e Which is the lightest baby?

f What is the range of their weights?

5 West Ham scored 2, 0, 2, 3 goals in four matches.

a What is the least number of goals scored in one match?

b What is the greatest number of goals scored in one match?

c What is the range of goals scored?

6 Arsenal scored 5, 4, 0, 2, 4 goals in five matches.

a What is the least number of goals scored in one match?

b What is the greatest number of goals scored in one match?

c What is the range of this data?

7 Aston Villa scored 0, 2, 1, 2 goals in four matches.

a What is the least number of goals scored in one match?

b What is the greatest number of goals scored in one match?

c What is the range of this data?

4 Misleading statistics

The Minister of Health and Social Security is presenting information during an election campaign. His charts show increases in spending on the National Health Service.

Do the charts give a fair picture of the figures?

Exercise 16:5

1 This graph shows baby Lisa's length.

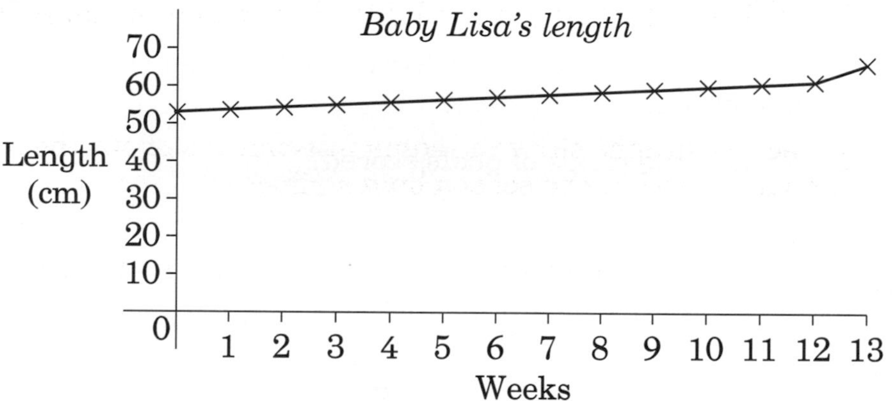

This graph also shows baby Lisa's length.

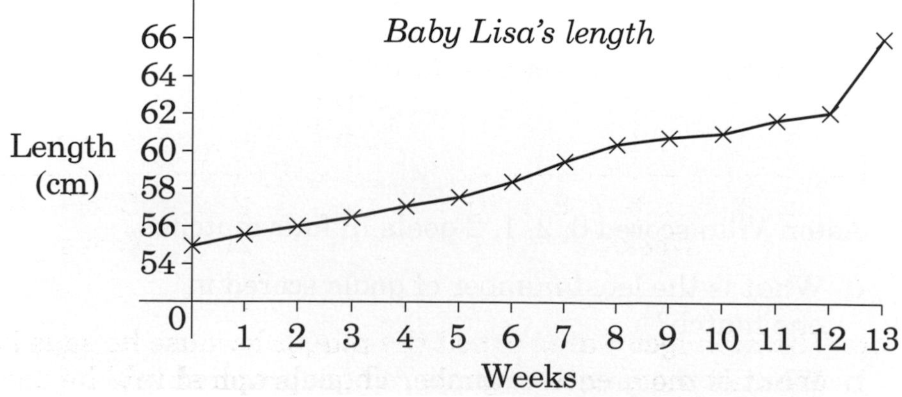

The graphs look different because the scales are different. Which graph makes it look like Lisa has grown the most?

2 These two graphs show the same information about baby Lisa's weight.

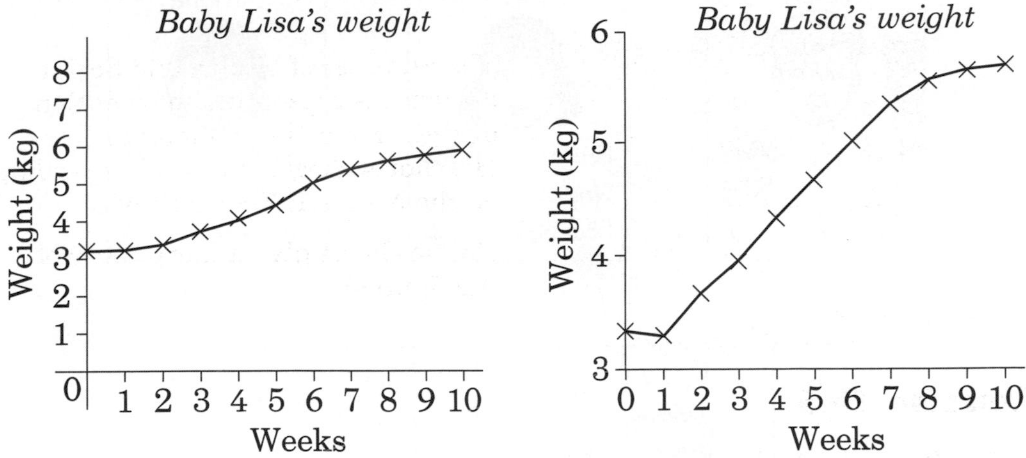

Which graph makes you think Lisa has put on the most weight?

3 These two graphs show the same information about the number of goals scored in one season by a striker.

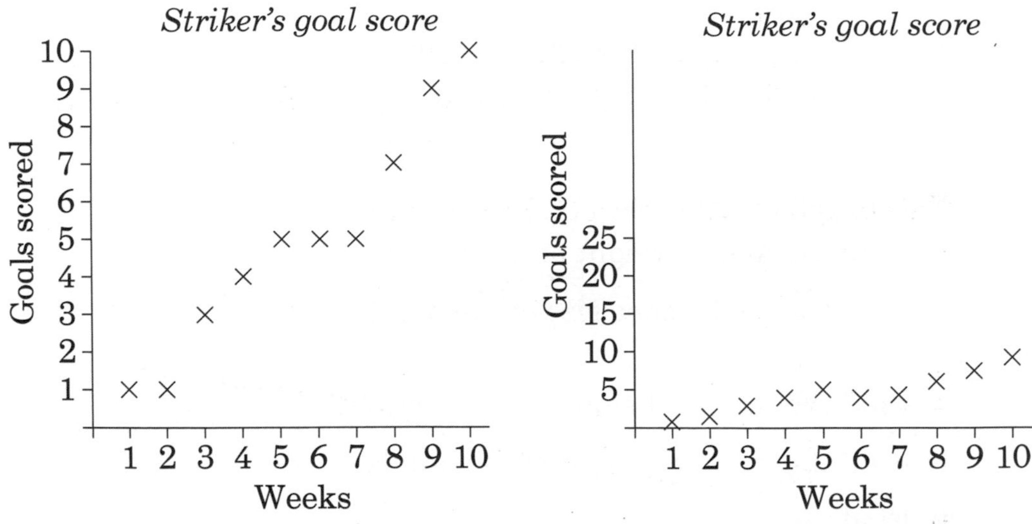

a The manager wants to sell the player because he says he does not score enough goals. Which graph should he use?

b The team captain wants to keep the player in the team. Which graph should he use?

1 Here are 5 children from Rosie's class.

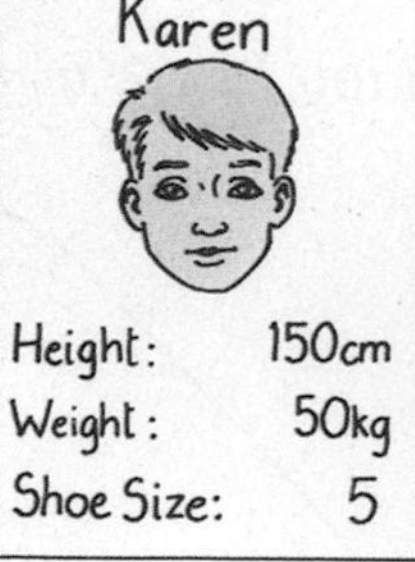

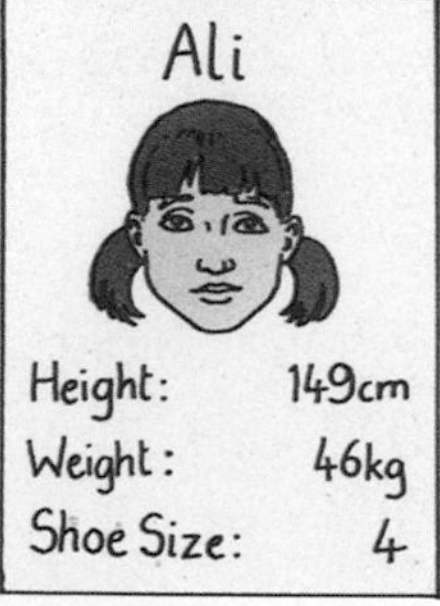

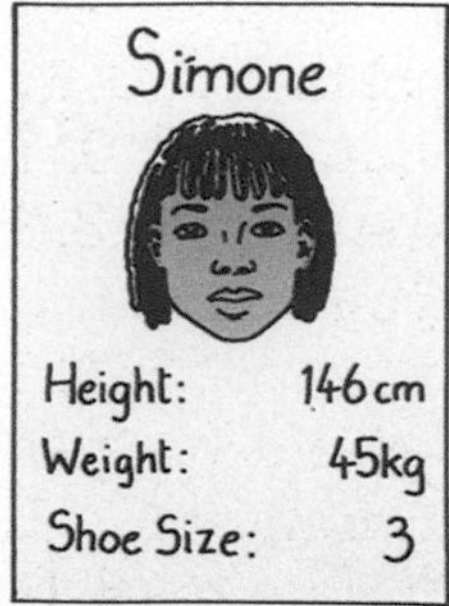

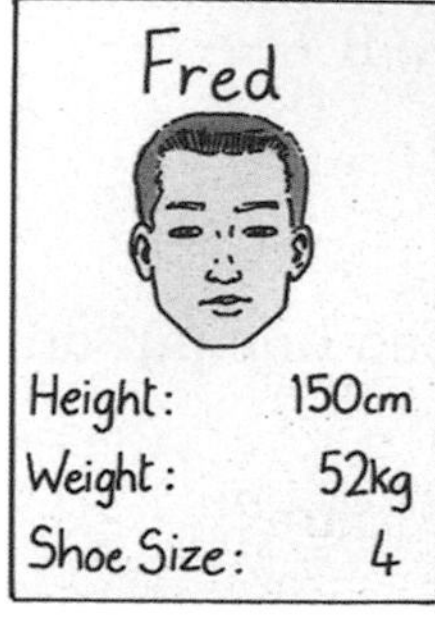

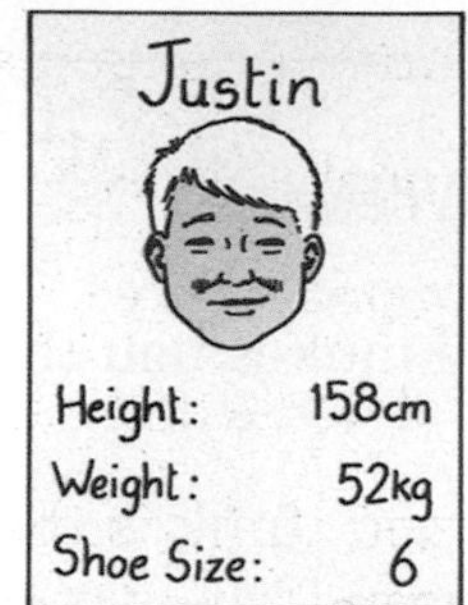

a Find the mean height of the 5 pupils.

b Find the mean weight of the pupils.

c Find the mean shoe size of the pupils.

d Find the modal height of the 5 pupils.

e Find the modal weight of the pupils.

f Find the modal shoe size of the pupils.

g Find the median height of the 5 pupils.

h Find the median weight of the pupils.

i Find the median shoe size of the pupils.

j Find the range of heights of the 5 pupils.

k Find the range of weights of the pupils.

l Find the range of shoe sizes of the pupils.

2 Here is some data about football teams and their goal scores.

Blackburn 0, 1, 3, 1, 7
Bolton 0, 0, 0, 1, 0
Coventry 2, 3, 1, 0, 1
Derby 1, 3, 4
Leicester 3, 1, 1, 0, 3
Sheffield Wednesday 0, 2, 3, 1, 1
Southampton 1, 0, 1, 0, 3
Tottenham 3, 0, 1, 0, 0

For each team find:

a the mean number of goals scored

b the modal goal score

c the median goal score

d the range of goals scored.

e Which do you think is the best team?
Give a reason for your answer.

3 Here is some data about pocket money.

Sarah £4	Peter £8	Angela £6
Georgia £7	Sammy £5	Sue £9
Emily £2	Lewis £5	Aaron £6

a What is the mean amount of pocket money?

b What is the modal amount of pocket money?

c What is the median amount of pocket money?

d What is the range of pocket money received?

Time 3

Match the clocks that say the same time.

1

2

3

4

5

6

7

8

Chapter 1 – Addition 1

Exercise 1

1 10 **2** 12 **3** 15 **4** 14 **5** 11 **6** 12 **7** 16 **8** 14

Exercise 2

1 16 **2** 19 **3** 19 **4** 19 **5** 17 **6** 20 **7** 16 **8** 17

Exercise 3

1 18 **2** 18 **3** 18 **4** 20 **5** 18 **6** 17 **7** 17 **8** 18

Exercise 4

1 20 **2** 19 **3** 19 **4** 18 **5** 19 **6** 20 **7** 19 **8** 20

Chapter 2 – Subtraction 1

Exercise 1

1 2 **2** 5 **3** 5 **4** 8 **5** 3 **6** 13 **7** 14 **8** 12 **9** 11 **10** 10

Exercise 2

1 14 **2** 11 **3** 13 **4** 21 **5** 22 **6** 27 **7** 13 **8** 14 **9** 12 **10** 13

Exercise 3

1 8 **2** 7 **3** 6 **4** 7 **5** 7 **6** 17 **7** 15 **8** 19 **9** 8 **10** 11

Exercise 4

1 4 **2** 4 **3** 9 **4** 8 **5** 18 **6** 18 **7** 15 **8** 19 **9** 18 **10** 26

Chapter 3 – Multiplication 1

Exercise 1

1 2 **2** 10 **3** 8 **4** 20 **5** 20 **6** 6 **7** 45 **8** 70 **9** 20 **10** 25 **11** 18 **12** 30 **13** 50 **14** 40 **15** 10 **16** 16

Exercise 2

1 40 **2** 60 **3** 5 **4** 40 **5** 14 **6** 4 **7** 20 **8** 10 **9** 15 **10** 12 **11** 90 **12** 2 **13** 20 **14** 20 **15** 100 **16** 18

Exercise 3

1 16 **2** 30 **3** 12 **4** 35 **5** 80 **6** 2 **7** 10 **8** 20 **9** 30 **10** 50 **11** 40 **12** 15 **13** 8 **14** 6 **15** 5 **16** 45

Exercise 4

1 35 **2** 18 **3** 8 **4** 30 **5** 40 **6** 25 **7** 4 **8** 20 **9** 14 **10** 12 **11** 60 **12** 70 **13** 10 **14** 40 **15** 100 **16** 2

Chapter 4 – Division 1

Exercise 1

1 1 **2** 6 **3** 2 **4** 7 **5** 3 **6** 8 **7** 4 **8** 9 **9** 5 **10** 10

Exercise 2

1 4 **2** 3 **3** 1 **4** 2 **5** 5 **6** 10 **7** 6 **8** 8 **9** 9 **10** 7

Exercise 3

1 2 **2** 5 **3** 4 **4** 8 **5** 10 **6** 9 **7** 6 **8** 3 **9** 7 **10** 1

Exercise 4

1 2 **2** 1 **3** 8 **4** 10 **5** 3 **6** 9 **7** 7 **8** 8 **9** 6 **10** 5

Chapter 5 – Money 1

Exercise 1

1 22p **2** 75p **3** 30p **4** 96p **5** £1.28 **6** £1.19 **7** £1.03 **8** £4.72 **9** £5.68 **10** £10.08

Exercise 2

1 Forty-two pence
2 Sixty-eight pence
3 Ninety-seven pence
4 One pound and one penny
5 One pound and thirty-eight pence
6 One pound and seventy-two pence
7 Three pounds and twenty-nine pence
8 Five pounds and sixty-one pence
9 Six pounds and ninety-three pence
10 Twenty pounds and five pence

Chapter 6 – Time 1

1 Three o'clock
2 Five o'clock
3 Half past eight
4 Half past one
5 Half past nine
6 A quarter past ten
7 A quarter past three
8 A quarter to seven

Chapter 7 – Addition 2

Exercise 1

1 79	**3** 38	**5** 39	**7** 64
2 38	**4** 47	**6** 47	**8** 39

Exercise 2

1 60	**3** 59	**5** 103	**7** 61
2 78	**4** 53	**6** 110	**8** 86

Exercise 3

1 61	**3** 64	**5** 44	**7** 100
2 69	**4** 42	**6** 64	**8** 108

Exercise 4

1 77	**3** 49	**5** 61	**7** 88
2 49	**4** 113	**6** 116	**8** 123

Chapter 8 – Subtraction 2

Exercise 1

1 12	**4** 8	**7** 48	**10** 26
2 8	**5** 33	**8** 22	
3 24	**6** 19	**9** 51	

Exercise 2

1 12	**4** 30	**7** 42	**10** 22
2 23	**5** 12	**8** 21	
3 32	**6** 21	**9** 50	

Exercise 3

1 45	**4** 79	**7** 38	**10** 28
2 57	**5** 39	**8** 18	
3 67	**6** 58	**9** 78	

Exercise 4

1 10	**4** 20	**7** 60	**10** 60
2 30	**5** 30	**8** 80	
3 20	**6** 30	**9** 50	

Chapter 9 – Multiplication 2

Exercise 1

1 30	**5** 90	**9** 80	**13** 70
2 50	**6** 300	**10** 600	**14** 900
3 700	**7** 400	**11** 20	**15** 1000
4 60	**8** 40	**12** 100	**16** 200

Exercise 2

1 500	**5** 90	**9** 600	**13** 300
2 10	**6** 30	**10** 60	**14** 80
3 200	**7** 700	**11** 500	**15** 400
4 40	**8** 20	**12** 70	**16** 800

Exercise 3

1 50	**5** 70	**9** 20	**13** 120
2 150	**6** 80	**10** 90	**14** 40
3 30	**7** 60	**11** 30	**15** 180
4 130	**8** 200	**12** 50	**16** 130

Exercise 4

1 80	**5** 1200	**9** 350	**13** 430
2 70	**6** 410	**10** 320	**14** 240
3 90	**7** 1700	**11** 210	**15** 470
4 140	**8** 320	**12** 160	**16** 510

Chapter 10 – Division 2

Exercise 1

1 1	**4** 2	**7** 5	**10** 6
2 7	**5** 8	**8** 9	
3 5	**6** 3	**9** 4	

Exercise 2

1 9	**4** 4	**7** 7	**10** 10
2 2	**5** 3	**8** 8	
3 1	**6** 6	**9** 5	

Exercise 3

1 10	**4** 9	**7** 1	**10** 5
2 4	**5** 3	**8** 2	
3 6	**6** 8	**9** 7	

Exercise 4

1 1	**4** 6	**7** 7	**10** 4
2 9	**5** 5	**8** 8	
3 3	**6** 2	**9** 10	

Chapter 11 – Money 2

Exercise 1

1 30p	**5** 70p	**9** £1.10
2 40p	**6** 90p	**10** £1.20
3 70p	**7** £1.00	
4 90p	**8** £1.10	

Exercise 2

1 20p	**5** 80p	**9** £1.20
2 40p	**6** 80p	**10** £1.60
3 30p	**7** 90p	
4 60p	**8** £1.00	

Exercise 3

1 30p	**5** 92p	**9** £1.54
2 31p	**6** 81p	**10** £1.26
3 43p	**7** £1.14	
4 57p	**8** £1.40	

Chapter 12 – Time 2

1 A quarter past three
2 A quarter to nine
3 Ten past seven
4 Twenty-five past three
5 Twenty past one
6 Five past ten
7 Five to nine
8 Twenty to seven

Chapter 13 – Multiplication 3

Exercise 1

1 6	**5** 12	**9** 10	**13** 18
2 24	**6** 20	**10** 8	**14** 18
3 15	**7** 8	**11** 18	**15** 24
4 16	**8** 30	**12** 28	**16** 36

Exercise 2

1 12	**5** 18	**9** 24	**13** 21
2 24	**6** 27	**10** 15	**14** 16
3 21	**7** 16	**11** 32	**15** 36
4 10	**8** 12	**12** 14	**16** 42

Exercise 3

1 48	**5** 20	**9** 12	**13** 45
2 15	**6** 18	**10** 30	**14** 40
3 54	**7** 60	**11** 12	**15** 16
4 14	**8** 25	**12** 28	**16** 35

Exercise 4

1 20	**5** 300	**9** 70	**13** 50
2 60	**6** 30	**10** 600	**14** 80
3 100	**7** 500	**11** 40	**15** 400
4 90	**8** 800	**12** 900	**16** 700

Chapter 14 – Division 3

Exercise 1

1 11	**4** 14	**7** 12	**10** 11
2 15	**5** 13	**8** 15	
3 12	**6** 14	**9** 13	

Exercise 2

1 21	**4** 22	**7** 31	**10** 33
2 34	**5** 43	**8** 23	
3 32	**6** 21	**9** 32	

Exercise 3

1 11	**4** 14	**7** 14	**10** 12
2 15	**5** 13	**8** 13	
3 12	**6** 11	**9** 15	

Exercise 4

1 16	**4** 51	**7** 18	**10** 56
2 111	**5** 31	**8** 47	
3 19	**6** 33	**9** 21	

Chapter 15 – Money 3

Exercise 1

1 20p	**5** 90p	**9** £1.80
2 20p	**6** 80p	**10** £1.70
3 40p	**7** 70p	
4 80p	**8** 90p	

Exercise 2

1 30p	**5** 40p	**9** £1.10
2 20p	**6** 80p	**10** £1.30
3 20p	**7** 10p	
4 40p	**8** £1.10	

Exercise 3

1 12p	**5** 71p	**9** £1.13
2 28p	**6** £1.08	**10** £1.54
3 46p	**7** £1.02	
4 14p	**8** £1.28	

Chapter 16 – Time 3

1 and **8**
2 and **4**
3 and **7**
5 and **6**